14 98

Messerschmitt BF 109

The Operational Record

Jerry Scutts

This edition first published in 1996 by Motorbooks International Publishers & Wholesalers, 729 Prospect Avenue, PO Box 1, Osceola, WI 54020 USA

© 1996 Jerry Scutts

Previously published by Airlife Publishing Ltd, Shrewsbury, England.

All rights reserved. With the exception of quoting brief passages for the purpose of review no part of this publication may be reproduced without prior written permission from the Publisher.

Motorbooks International is a certified trademark, registered with the United States Patent Office.

The information in this book is true and complete to the best of our knowledge. All recommendations are made without any guarantee on the part of the author or publisher, who also disclaim any liability incurred in connection with the use of this data or specific details.

We recognize that some words, model names and designations, for example, mentioned herein are the property of the trademark holder. We use them for identification purposes only. This is not an official publication.

Motorbooks International books are also available at discounts in bulk quantity for industrial or sales-promotional use. For details write to Special Sales Manager at the Publisher's address.

Library of Congress Cataloging-in-Publication Data Available.

ISBN 0-7603-0262-6

Typeset by Phoenix Typesetting, Ilkley, West Yorkshire.

Printed and bound in Great Britain.

Acknowledgements

The author makes due acknowledgement to the following published works dealing solely or in part with the history of the Bf 109. While this must be only a partial list, these titles are among those that contain the most accurate data. *The Messerschmitt Bf 109 F to K* by Joachen Prien & Peter Rodeike; *The Legion Condor* by Karl Ries & Hans Ring; *The Messerschmitt Bf 109 in Italian Service* by Angelo D'Amico; *The German Fighter since 1915* by Rudiger Kosin; *Sea Eagles* by Francis Marshall; *Six Months to Oblivion* by Werner Gerbig; *History of the German Night Fighter Force* by Gebherd Aders; *The Rise and Fall of the German Air Force* edited by W H Tatum IV & E J Hoffschmidt and *The Last Year of the Luftwaffe* by Alfred Price. The author also passes on his thanks for photographs supplied by among others, Peter Petrick, Eddie Creek, James V Crow, Bruce Robertson, Philip Jarrett, Harry Holmes, Brian Marsh, Dick Ward, Dave Howley and Merle Olmsted.

Contents

Introduction		xi
Chapter 1	**Genesis**	**1**
	Berta, Clara and Dora	9
	First of the Many	13
	Exports	16
Chapter 2	**First Blood**	**19**
	The Decisive Year	23
	Aragon	23
Chapter 3	**East–West Triumph**	**28**
	Against the RAF	31
	Improved Emils	33
	Battle for France	34
Chapter 4	**Channel Clash**	**40**
	Testing the Friedrich	43
	Adler Tag	47
	Enter the F	52
	Jabos	53
	Last of the Emils	53
Chapter 5	**Mediterranean Debut**	**55**
	Yugoslavia and Greece	61
	Crusader	64
	Torch	67
Chapter 6	**New Threats**	**70**
	High flying Gustav	73
	Night Moves	77
	Swiss Adventure	78
Chapter 7	**Russian Roulette**	**79**
	Stalemate	87

Chapter 8	**Three Front Disaster**	**91**
	Improved Gustav	96
	Invasion Front	103
	Fast Reconnaissance	104
Chapter 9	**East-West Debacle**	**109**
	Ray of Hope	111
	Experimental 109s	116
Chapter 10	**Italian Swansong**	**118**
	German Coup Attempt	122
Chapter 11	**Defeat**	**126**
	Aftermath	132
Chapter 12	**Fading Away**	**134**
	Spain's 'Fat Pigeons'	136
Index		143

Introduction

One of the truly great military aircraft of all time, the Messerschmitt Bf 109 alone exemplified the wartime *Luftwaffe* in the minds of many people. It was fully tested and became operational some time before the start of that terrible conflict and was still going strong when Germany surrendered in 1945. It was built in larger numbers than any other type committed to combat by a Western nation and its service record far surpassed that of any other fighter, for the total number of aerial victories scored by its pilots is very unlikely ever to be even approached, let alone equalled or beaten.

Known variously by a host of epithets, most of which emphasised its tendency to be a demanding beast, sometimes dangerous and occasionally lethal to the unwary, the '109' nevertheless forged an enormous affection in the hearts of German and Axis fighter pilots and at least in its early wartime career, a healthy respect from Allied aircrew. Its legendary reputation endured long past retirement.

Although technically equalled quite early on in its career, not only by the products of nations at war with Germany but by the Focke-Wulf 190, the 109 soldiered on, a proven workhorse, familiar, invariably available and, with a good pilot at the controls, often equal to the challenge that the burgeoning aerial might of the Allies imposed upon it.

Anyone reading a range of literature on the German fighter force in WWII cannot have failed to notice the nicknames the Bf 109 seems to have picked up during its career. These began with the more or less 'official' terminology for model designations derived from the German phonetic radio code: limited use was made of 'Anton' because the first Bf 109s were not A models as such, but *Berta, Clara,* (*Caesar* in some cases, apparently), *Dora, Emil, Friedrich, Gustav* and *Toni* were frequently used. These have passed into popular usage to readily identify the Bf 109B, C, D, E, F, G and T with the less common *Karl* or *Kurfürst* applied to the Bf 109K. Then there was the service slang.

To German pilots, any aircraft was invariably a 'crate' and *Beule* (bump) is a familiar enough term for the *Gustav*, with its prominent bulged gun breech covers. *Kannonboot* (gunboat) was a term widely applied – again mostly to *Gustavs* – to identify aircraft equipped with the twin gondola-mounted underwing cannon. But the 109 was also known as a *Badewanne* (bath tub), *Kahn* (canoe) and *Kiste* (box). Less obvious perhaps was an oblique reference to the weakness of its undercarriage by likening it to a *Molle*, a drinking glass found in Berlin restaurants – and therefore breakable!

How common these terms were throughout the *Luftwaffe* or in any individual unit is really a question only an ex-Bf 109 *Jagdflieger* (fighter pilot) could answer with any certainty, as useage probably varied considerably. But the fact remains that in the flyers' lexicon, which was usually leavened with a degree of black humour, the more detrimental the nickname, the higher the object of seeming derision was actually regarded. So it was with the *ein hundert neun*.

Conceived and matured during the early years of National Socialist government in Germany, the Bf 109 was to become the backbone of the new *Luftwaffe*, the air force that would restore national pride by force of arms and finally wipe out the perceived shame of the Treaty of Versailles, the prognostications of which so humbled the country after WWI. Before undertaking such a monumental venture, the cornerstone of which was a strong air force, Germany required a modern monoplane fighter – and the Messerschmitt Bf 109 filled that requirement more successfully than anyone could have imagined.

JERRY SCUTTS

Chapter 1
Genesis

During the 1930s, Germany was far from alone in seeking to modernise her military air strength. The biplane's day was clearly waxing and something of a race developed among the world's most powerful nations to replace the aeroplanes that were little improved over those of the first war, with the new monoplanes. In the event it was the Soviet Union that gained the distinction of forming the first operational units equipped with monoplane fighters in the shape of the Polikarpov I-16 of 1933.

Germany was not far behind, at least in laying the foundation of a modern industry capable of a similar achievement; among a hard core of brilliant individuals who were eventually to give the *Luftwaffe* a range of modern combat aircraft was Wilhelm Emil Messerschmitt, universally known as Willy. Having cut his teeth on glider design before WWI in company with another youngster, Friedrich Harth, Messerschmitt completed a short period of military service during the war and by the armistice of 1918, had entered the *Technische Hochschule* in Munich. Meanwhile Harth had joined the *Bayerische Flugzeug Werke* (*BFW* or Bavarian Aircraft Company).

Although powered flying was circumscribed in a defeated Germany, gliders could be flown without restriction and when designer Oskar Ursinus opened a flying centre in the Rhoen area, Messerschmitt and Harth were the first to respond to the call for candidates. The young engineers now found an outlet for their talents and enthusiasm for flying and by 1921 Willy Messerschmitt had progressed to his tenth glider design. Unfortunately Harth suffered a bad crash that year which left him permanently disabled. Parting company with Harth in 1923, Messerschmitt remained at the technical school but founded his own flying group at Bamberg, northern Bavaria.

An able aeronautical design student, Messerschmitt became the first individual to be allowed to build a glider as part of his thesis. This successful project helped him to pass the necessary examinations which put him firmly on the road to becoming a fully qualified engineer and designer.

Collaboration with Theo Croneiss, the director of *Sportflug GmBH*, inspired Messerschmitt to more ambitious projects. Messerschmitt put his design stamp on 14 gliders before graduating to his first powered machine, the S 15. Whereas all the gliders had been prefixed by the letter 'S' for *Segelflugzeug* (Sailplane), the powered aircraft were prefixed by 'M' for Messerschmitt and included the M 17 light aircraft of 1926, with the M 18 airliner following a year later. Working capital to build these aircraft was provided both by Messerschmitt's own company and *Sportflug* and the export success of the M 18 led to further expansion with the formation of *Flugzeugbau Messerschmitt Bamberg*.

Seeking further funding in the form of a subsidy from the Bavarian government, Messerschmitt was told that available finance was already allocated to the *BFW* company. It was suggested however, that the Messerschmitt concern might merge with *BFW*, with the latter undertaking to build solely indigenous designs. In his turn, Messerschmitt would give *BFW* first priority in allocating such work. This was agreed and on 8 September 1927 Willy Messerschmitt became chief designer of the new organisation with his headquarters at Augsburg.

Finance however continued to be difficult to come by. But Messerschmitt and a wealthy friend, Baroness Lilly von Michel-Raulino, respectively acquired 12.5 per cent and 87.5 per cent of the

One aircraft that greatly helped Messerschmitt's pre-war finances was the Bf 108 'Taifun'. This example is believed to have been used by German Embassy staff in the UK.

Bavarian government shares in *BFW*. Messerschmitt became joint managing director with Fritz Hille, but while the company undoubtedly had expertise, it could not sell enough airframes to keep viable and *Bayerische Flugzeug Werke* filed for bankruptcy in June 1931.

Desperately casting around for a way to keep at least a nucleus of the company solvent, Messerschmitt sold his car and managed to keep his design team together and secure his patents. A deal with a Romanian concern for the licence-production of his M 23b lightplane followed an acrimonious period during which *Lufthansa* cancelled then reinstated an order for the M 28 airliner. A small group of workers could thus be retained under the aegis of a new private company, *Messerschmitt-Flugzeug GmBH*.

The M 23 was a commercial success for Messerschmitt, who not only managed to pay off most of his creditors but had made a firm friend of Rudolf Hess, deputy leader of the National Socialist party. When Adolf Hitler came to power in 1933, this contact undoubtedly helped Messerschmitt to remain in business. Indeed, he was favoured by the party to the extent of having his firm form the basis of a larger, state-owned organisation, *Bayerische Flugzeug Werke A.G.* Having discharged its bankruptcy in April 1933, *BFW* was able to resume business as an aircraft manufacturer, albeit on a modest scale.

Messerschmitt's forte had and continued to be light, efficient sport aircraft that owed much to his sailplane experience by exhibiting an amazingly low power to weight ratio. Never one for hanging things on airframes if they were not absolutely necessary, Messerschmitt gained a powerful international reputation as his creations won international flying competitions. Hess was amongst his customers for the M 35 and after the M 36, Messerschmitt began work on the M 37 four-seat cabin monoplane, which eventually became the Bf 108.

The new government of Germany was however, a mixed blessing for Messerschmitt, who had made a few enemies in his rise to power, among them Erhard Milch. As a director of *Lufthansa,* Milch had had cause to question the design integrity of *BFW*'s aircraft following a series of fatal crashes. As the Secretary of State for Air and deputy to Hermann Göring, Reich Minister for Aviation, Milch was in a position to make life difficult for Messerschmitt and it was not long before his resentment worked through. An edict to the effect that *BFW* would henceforth manufacture the products of other companies had to be complied with. Messerschmitt also fell foul of Ernst Heinkel, an industrial rival.

Luckily, Messerschmitt was able to secure advance orders from abroad, initially for the M 37; an increasing order book for this aircraft provided a firmer industrial base, which in turn, helped provide the finance to develop the Bf 109 single-seat fighter. Ably assisted by designer Robert Lusser, who joined *BFW* in 1933, Messerschmitt turned his attention to the future requirements of the *Luftwaffe*, then an open secret in Germany. Apart from two minor, unsuccessful biplanes, *BFW* lacked any experience in the design of military aircraft, a fact that his rivals were only too pleased to stress. Messerschmitt however responded positively to a 1934 requirement issued by the *Technisches Amt* (Technical Office) of the *Reichsluftfahrtministerium* (*RLM* – Air Ministry) for a new tactical single-seat fighter for the *Luftwaffe*.

Among the attributes that the *RLM C-Amt* required in the winning design were: a maximum speed of 400 km/h at 6,000 m (248 mph at 19,680 ft); a duration of one and a half hours, a service ceiling of 10,000 m (32,800 ft) and an armament of either two fixed machine guns each with

Pictured on a test flight over Augsberg in 1936, the Bf 109V-3 was subsequently evaluated in Spain.

At least 20 Bf 109s carried pre-war civil registrations, among them the V5, D-IEKS *(Robertson)*.

1,000 rounds of ammunition or a single 'machine cannon' with 200 rounds. Both day and night aerial combat was envisaged and for the latter role, the specification included the sentence: 'It must be possible to install appropriate equipment for night flying.'

Messerschmitt hardly had the field of fighter design to himself, as Heinkel, Arado and Focke-Wulf responded enthusiastically to the competition, the winner of which could expect a highly lucrative contract. Each company tended to follow different design approaches, Focke-Wulf being particularly impressed with the parasol wing for its Fw 159, while Arado, although favouring the 'modern trend' towards low-wing cantilever monoplanes, retained an outmoded, fixed undercarriage for the Ar 80. Heinkel represented *BFW*'s closest rival with the He 112, an elegant low wing monoplane which in early form, featured an open cockpit.

Messerschmitt held only a faint hope that his new fighter might be chosen for series production – he had apparently been told in no uncertain terms that even if he produced an exceptional design, it would find little favour in official circles. Only reluctantly had the *RLM* (thanks to Milch) even agreed to allow Messerschmitt to enter the competition.

In order to concentrate the best resources on the fighter project, Messerschmitt created a new design team led by Lusser as chief design engineer and including Hubert Bauer, workshop manager and head of experimental construction. They strove to create the smallest possible fighter airframe around the most powerful engine then available, drawing on the layout of both Messerschmitt's M 29 lightplane of 1932 and the M 37.

The company meanwhile concerned itself with the impending first flight of the Bf 108. This singularly significant event for Willy Messerschmitt took place on 13 June 1934. By then, work had begun on the Bf 109 mock-up in readiness for inspection by *RLM* officials that October, both Heinkel and Arado mock-ups having been viewed earlier in the year.

A subsequent meeting of the design competition committee was chaired by Milch, who confirmed that the fighter ordered for the *Luftwaffe* was to be a cantilever low-wing monoplane with a retractable landing gear. Part of the discussion centred on technical aspects of aircraft design including wing loading and it was stated that whereas Heinkel and Arado had been restricted to 100 kg/m² (16 lb/sq ft), Messerschmitt was allowed to go to 125 kg/m² (20 lb/sq ft). This was of upmost importance to the lightweight fighter Messerschmitt planned; in order to maintain aileron effectiveness at low speeds, slotted flaps would be used, mated to a low-set cantilever single-spar wing with a torsionally-stiff leading edge incorporating a moveable slat. All control surfaces were also fabric covered to save weight.

While *BFW* proceeded to build the prototype Bf 109, Milch had further voiced his reluctance for Messerschmitt's continued participation, saying that, 'Nothing much will come of it but as a pike in the carp pool Messerschmitt may be quite good.' This type of comment had become familiar enough to Messerschmitt although looked at in a positive light, it was a veiled testimonial to an expertise other firms would be hard put to match.

In the meantime *BFW* initiated production of the Me 108, six examples being completed before the end of 1934. By then some design work on the

With pole-mounted instrumentation on the nose and a device attached to the fin, this machine is understood to have been the Bf 109 V6, D-IHHB.

Bf 110, the so-called 'heavy fighter', or *Zerstörer* (destroyer) had also been initiated at Augsburg.

On 1 March 1935, Hitler removed the veil of secrecy that had hitherto surrounded the existence of the *Luftwaffe* by proclaiming that the new air force would no longer operate under cover with all the problems that that had entailed. *Enttarnung* or 'de-camouflaging', meant that aircraft orders could openly be placed with manufacturers who in turn could expand their production plants wherever necessary, assisted by generous *Reichsbank* financing.

By the early spring of 1935 the prototype Bf 109, the V1, allocated the civil registration D-IABI, was all but ready for its first flight. One major problem was that its intended powerplant, the Junkers Jumo 210A, was not yet available. Delivering 610 hp, the Jumo engine promised adequate horsepower but in order to fly his prototype without further delay, Messerschmitt oversaw the installation of the more powerful 695 hp Rolls-Royce Kestrel V.

Flugkapitan Hans-Dietrich 'Bubu' Knotzsch duly took the Bf 109 V1 up for its maiden flight from *BFW*'s Augsburg-Haunstetten airfield on 28 May 1935. The flight was satisfactory, with few problems anticipated, Knotzsch enthusing over the prototype's excellent handling characteristics. There followed some four months of factory testing before the machine was moved to the *E-Stelle* (experimental establishment) at Rechlin on 15 October for *Oberkommando der Luftwaffe (OKL)* (*Luftwaffe* High Command) evaluation, a programme that included comparative tests with the Ar 80 V1, the Fw 159 V1 and the He 112 V1.

Clearly, only the Heinkel and Messerschmitt designs were configured to meet the requirement previously decided upon for the *Luftwaffe*'s new fighter – in fact so closely were they matched in performance that the *RLM* ordered ten further pre-production examples of each aircraft. This caused some surprise in official circles, as Messerschmitt had not been thought of as a serious contender for the fighter contract.

BFW completed the Bf 109 V2 (D-IDUE) with the Junkers Jumo engine installed and flew it for the first time on 21 January 1936. The first prototype had had no armament provision and as was more or less standard international practice, guns would only be fitted once some basic performance figures and flight characteristic data were to hand. In the case of the Bf 109, the V2 was the first example to be armed, two MG 17 7.9-mm machine guns being installed above the engine to fire, via interrupter gear, through the airscrew disc from troughs set into the top of the cowling.

The guns were aimed and fired by a *Revi* reflector sight which was progressively improved in detail during the Bf 109's production span, but it remained basically the same. The guns were fired electrically by depression of a trigger on the control column by the pilot's index finger, after the weapons had been pneumatically cocked.

A gun set to fire between the 'V' configuration cylinder banks, with the shells exiting through the centre of the airscrew spinner, had been envisaged for the Bf 109 from the earliest design stage and such a weapon was installed for the first time on the V-4, D-IOQY. A 7.9-mm machine gun rather than the intended 20-mm MG FF cannon, this weapon nevertheless provided useful weight and balance data. In the event, it was to be some time before the weight and vibration problems associated with a centreline cannon installation were overcome.

In February 1936, the final comparison flights between the Bf 109 and He 112 were conducted. The new fighters were flown by Ernst Udet, one of Germany's most experienced pilots as well as *Major* Robert Ritter von Greim and the Rechlin test pilots including *Dipl-Ing* Carl Francke. Hermann Wurster, who had only recently joined *BFW*, completed the flight test team. The Bf 109 V2 and the He 112 V2 were selected for the final performance assessment at Travemunde, the Messerschmitt being flown from there on 26 and 27 February and 2 March.

Wurster's demonstration of the Bf 109 V2 was masterly and all the more commendable considering that he had only flown the fighter fifteen times prior to this all-important official evaluation. His programme included a high speed dive from altitude and a demanding spin test, with a low altitude recovery. The impressive demonstration was in stark contrast to the ill-luck that befell the Heinkel fighter. Gerhard Nitschke experienced the ever-present hazards of test flying and the ultimate humiliation of losing a prototype when the He 112 could not be recovered from a spin. Nitschke survived, but the aircraft was totally destroyed.

Other aircraft have successfully weathered a prototype crash early in their careers and the demise of the He 112 V2 in no way prejudiced Rechlin's final evaluation between it and the Bf 109 V2 – indeed it was still far from certain at that stage, which fighter would be chosen for full series production. The criticisms of Messerschmitt's design included the weak undercarriage (which had unfortunately been publically demonstrated when the V1 suffered a ground loop at Rechlin), which was partly due to an innovative method of attaching both mainwheel oleo legs to the fuselage rather than the wing, necessitating a very narrow track.

Along with the test pilots, Ernst Udet flew both new fighters and although he voiced some enthusiasm for the He 112, he selected the Bf 109 to be the standard *Luftwaffe* fighter. In the interim, Messerschmitt and Milch appear to have patched up their differences and the relationship resulted in a steady flow of contracts for the Augsburg concern.

Among the positive attributes of Willy Messerschmitt's brainchild was that it was estimated to be far less demanding to build than its rival. The Bf 109's design had considerable mass production potential 'built in', particularly in regard to the wing-fuselage join, which required only three pins. The fact that prior to this stage completed fuselages could be moved around on their mainwheels saved factory floor space and reduced assembly time.

In reaching its decision in favour of Messerschmitt, it is almost certain that the *RLM Techniches Amt* also recognised that in contrast to the Bf 109 which had no awkward curves, shaping the elliptical wing structure of the He 112 could be considerably time- and labour-intensive. Few production line bottlenecks were envisaged for the Bf 109, a factor that was to take on an increasing importance. The simplicity of an all-metal stressed skin, flush-riveted to the main structure and incorporating a typical Messerschmitt single-spar wing, was to be one of the major assets in building the Bf 109.

Ease of installation was also reflected in the design of the engine bay and by opting for an inverted-vee cylinder bank layout, Messerschmitt engineers ensured that the Bf 109 pilot's forward view was not obscured any more than was necessary. This was an international problem faced by numerous designers of fighters with a steep fuselage ground angle caused by adopting a tailwheel configuration, the generally accepted approach at that time.

Even though the Junkers Jumo had been the originally-specified powerplant, the Daimler Benz company then had the first examples of the powerful new DB 600/601 engine in the final testing stage. Fitting the Daimler Benz engine would give the Bf 109 a significant increase in output (to 1,300 hp) and consequently improved performance. The DB 600 had marginally greater overall dimensions than the Jumo, but the design of the 109's forward fuselage was such that very little modification was required to fit an alternative powerplant and the centre of gravity was not adversely affected.

The Bf 109 underwent an extensive test programme for engines and equipment and up to the outbreak of war, Messerschmitt completed a series of 20 aircraft that were given *Versuchs* (Experimental) numbers and civil registrations. These were in some cases 'one offs' but other V-numbered examples began life as standard production airframes which were retained for company testing. Early on, the diverse tests included exploring the potential of the new fighter as a carrier-borne interceptor.

In the latter half of 1936, production of the Jumo-powered Bf 109 as the B-1 was initiated at the main *BFW* plant at Augsburg, one example becoming the V-17a (D-IKAC), used in the Bf 109T carrier-suitability programme. The outwardly similar B-2 followed in production soon afterwards and it became clear that with Messerschmitt's other military aircraft commitments, space at Augsburg was rapidly running out.

Messerschmitt was consequently to bring other production centres into the Bf 109 programme, including a second *BFW* facility at Regensburg which opened in 1937, the *Wiener Neustädter Flugzeugwerke* (WNF) facility in Austria, Erla at

Leipzig and Arado's Warnemunde plant. Eventually, Bf 109s were also assembled at Focke-Wulf's Bremen plant, by Fieseler at Kassel and by Dornier and Ago at Oschersleben. These satellite plants, rather than Augsburg, built a significant proportion of the total Bf 109s produced before and during the war.

Although the Bf 109 aroused considerable public interest when the V2 was briefly demonstrated at the 1936 Olympics in Berlin, Messerschmitt (who was appointed a Professorship in 1937) wished to prove how well his protégé could compete in a flying competition. He consequently put together a strong team for the fourth International Flying Meeting held at Zürich-Dubendorf Airport between 23 July and 1 August the following year. Three Bf 109 prototypes, the V7, (Zürich competition No 4), V10 (No 6) and V13 (No unknown), were accompanied to Switzerland by two production B-1s and a B-2.

Among the pilots who flew the Bf 109 at the Zürich meeting was *Maggiore* (Major) Aldo Remondino, head of Italy's national aerobatic team. Although it was not envisaged at the time, the Italians would later be closely associated with the Messerschmitt fighter and Remondino gained the distinction of being among the very first, if not the first, foreign pilot to fly the Bf 109.

Despite Ernst Udet accidentally writing off the V10, *Major* Hans Seidemann won the 228-mile Circuit of the Alps event in the Bf 109B-2 at an average speed of 241 mph, taking 56 minutes, 47 seconds to complete the course. Carl Franke flew the V13 into first place in the dive and climb competition, reaching 9,840 ft and diving to 500 ft in 2 min 5.7 sec.

The speed event was also won by the B-2, while

Bloodied in Spain, the Bf 109B-1s began the domination of the Republicans for Franco's Nationalists. This was the ninth B-1 to be delivered.

the team race resulted in yet another German victory, the Messerschmitts being piloted by *Hauptmann* (*Hptm*) Restemeier and *Leutnant* (*Lt*)s Trautloft and Schleif at an average speed of 233.5 mph. These results represented a resounding success for Germany's new fighter, the aviation industry in general and the country, and they greatly impressed the foreign visitors and pressmen.

On 11 November 1937 the Bf 109 V13 (D-IPKY) with Hermann Wurster at the controls, furthered German prestige by capturing the World Speed Record for Landplanes. Under *Fédération Aéronautique Internationale* (FAI) regulations, Wurster covered the 1.86-mile circuit twice in each direction at an altitude not exceeding 245 feet. His speed of 379.38 mph was enough to take the record, the first time that a German pilot flying an indigenous aircraft had done so. The Bf 109 was not out to capture the Absolute Air Speed Record, this being held by a seaplane, the Italian Macchi-Castoldi C 72 which had flown at 440 mph, a speed almost out of reach of a landplane at that time.

A little over a year after the Bf 109's success in Switzerland, on 11 July 1938, the *Bayerische Flugzeugwerke A.G.* was renamed *Messerschmitt A.G.* The original designation 'Bf' was retained in company records for the 109, although all other subsequent Messerschmitt designs took the abbreviated prefix 'Me'. Inevitably, the 109 was widely referred to as 'Me 109' but such tended to appear only in unofficial sources.

Berta, Clara and Dora

As the first production model of the Bf 109, the B-1 was powered by a Jumo 210Da engine of 680 hp and carried the modest armament of two fuselage mounted 7.9-mm machine guns. It was otherwise distinguished by a capped propeller spinner – a necessary step to reduce drag when the intended installation of an engine-mounted 20-mm cannon proved impracticable. The airscrew itself was manufactured by the Schwarz concern and was a fixed-pitch wooden unit with two blades. A fixed-pitch propeller imposed some performance penalties and work progressed to introduce variable pitch capability as soon as practicable.

Although a number of major improvements were made to the Bf 109B-2, it remained outwardly similar to the B-1 and was armed with three MG 17 7.9-mm machine guns, two in fuselage troughs and one positioned to fire through the airscrew spinner. With its wide central orifice, this spinner had been designed for aircraft fitted with an engine-mounted cannon and no further modification was of course necessary for a smaller-calibre machine gun, which easily fitted the blast tube.

From the pilot's viewpoint one of the most significant changes made to the Bf 109 B-2 was the fitting of a VDM variable pitch airscrew. This materially helped boost performance to 289 mph at 13,100 ft. Very rapidly introduced to combat in Spain, the B-2 enabled Condor Legion pilots to fight on more equal terms with Republican fighters, particularly the nimble and well-armed Polikarpov I-15 and I-16. The early 109s went

The distinctive faired cuffs of the Schwarz fixed-pitch wooden airscrew of the Bf 109B-1.

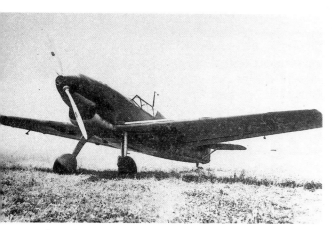

Significantly different in appearance and performance, the variable pitch prop of the Bf 109B-2 was a natural evolutionary step.

through a number of significant changes, not the least of which were aimed at boosting the armament but for some time, no significant increase in firepower was achieved. The Bf 109C-0 had two MG 17s in the now-standard 109 position in the upper cowling troughs and two in wing bays outboard of the undercarriage legs. Among other improvements made to the Bf 109C was the installation of a Jumo 210Ga engine of 650 hp, which dispensed with a carburettor system in favour of direct fuel injection. This enabled the pilot to perform all aerobatic manoeuvres without fear of the engine suddenly cutting out, particularly under conditions of negative G, as could happen with a float carburettor.

The Bf 109C-1 attained a top speed of 273 mph and had a service ceiling of 32,808 ft. Production was modest, with only about 50 being built between 1937 and 1938. Some C-1s were retrospectively fitted with the MG FF cannon in place of the wing-mounted MG 17s, the cannon being positioned further outboard than the

Ground crews of 2./JG 71, forerunner of JG 51, honing their refuelling and rearming technique on a Bf 109D-1 shortly before the war. (*Smithsonian Institution*)

Dispersed in a Spanish olive grove, *Jagdgruppe* 88 Bf 109D-1s coded '6-60' and '6-56' await the next call to arms. (*R L Ward*)

Strikingly marked on their predominantly dark green finish, the early Bf 109s made an impressive show of military might. These C models are believed to have been part of *JGr* 102 in 1939.

machine guns and in this form the aircraft was re-designated the C-3.

The Jumo engine also powered the Bf 109D-0 and the D-1, the development of both owing much to data accumulated during testing of the V10, 11, 12 and 13. Three variants (D-1, D-2 and D-3) were planned but in the event, all 650 airframes produced were designated as C-1s before BFW tooled up to build the more powerful and combat-capable 109E in 1938. The armament of the D-1 was the same as that of the D-1 and performance closely matched that of the earlier Jumo-engined sub type, the B-1. It appears that on the assembly lines the D-1 was not immediately supplanted by the *Emil*, the last *Dora* not being completed until 1939.

By that time the burgeoning *Luftwaffe* fighter force at home had received enough Bf 109s to replace or supplement biplane fighters in all 20 of the pre-war *Jadgeschwadern*. All these units (identified by three-digit numbers in the 200 range) would be absorbed under the new designations shortly to be adopted, as would seven *Jagdgeschwadern* with two-digit numbers and a couple of hybrid formations that belonged to neither of the above convenient groupings, namely *Jagdgruppen* 186 and 136.

This was the situation when in September 1938 Hitler's occupation of the Sudetenland area of Czechoslovakia precipitated the so-called Munich crisis and brought Germany that much closer to armed conflict. The number of Bf 109s then in *Luftwaffe* service (583 examples of all models) was widely believed – at least inside Germany – not to have been nearly enough with which to wage war on a strong neighbouring country. But the propaganda machine had done its work well and no military challenge to German territorial claims was forthcoming. Internationally, the *Luftwaffe*'s strength was

Switzerland's long association with the Bf 109 began with ten D-1s, this being the first example delivered. (MAP)

believed to be far greater than it actually was – but having 500 plus modern fighters in service was in fact substantially better than most other European countries had then been able to achieve.

A handful of *Luftwaffe* officers were however, aware of the possibility that in the event of war, German targets could indeed be bombed, not only by day, but by night. It will be recalled that the original requirement from which the Bf 109 stemmed contained the rider that the aircraft selected as the new *Luftwaffe* fighter should be capable of accommodating 'equipment for night flying'. And while this was undoubtedly a very minor adjunct to an all-encompassing specification, particularly as the hinted-at equipment did not then exist, the early 109s did participate in some early night fighting tests.

Experiments to explore the feasibility of destroying enemy night bombers by a combination of standard service interceptor fighters, searchlights and *Flak* (*Flugzeugabwehr-Kannonen*) – anti-aircraft guns – were begun as early as May 1936. These proved that such nocturnal sorties could result in successful interceptions and later, units flying the Bf 109C and D were involved. Despite the fact that night fighting was afforded a low priority, the idea was kept alive.

Eight *Jagdgeschwadern* were allowed to form night fighter *Staffeln* in 1939, yet these had reverted to standard day fighter duties by August, mere weeks before the outbreak of WWII brought the possibility of night bomber attacks on Germany that much closer. Then, night fighting was resurrected, albeit on a limited basis.

First of the Many

With a substantial number of test and operational flying hours having been accumulated by the prototypes as well as operational B, C and D models, the revisions made to the basic design to create the Bf 109E were based on sound data, much of it acquired under actual combat conditions in Spain. These changes were to result in an excellent, all-round fighter, the equal of any other warplane of similar configuration in existence anywhere in the world.

With its new radiator system, revised engine cowlings and three bladed propeller, the Bf 109E-1 was significantly different to its predecessors. This is an early example, almost certainly the V15. (*via Robertson*)

The Bf 109E first flew late in 1938, powered by the 1,100 hp DB 601A engine with direct fuel injection and having been modified, in comparison with the Jumo-engined variants, to improve the cooling system and reduce drag. By replacing the Jumo powerplant's large under-nose ducted radiator with a new, three-element system, a slimmer, less drag-inducing intake fairing was

A number of Bf 109 *Versuchs* airframes were fitted with the DB 601 engine and some were apparently issued to *Luftwaffe* training schools. Two examples are seen in this view. *(via Robertson)*

possible; completing the new system were two radiators located under the inboard section of the wings to give the *Emil* a design characteristic that distinguished it from all previous production models.

A three-blade VDM airscrew with full variable pitch was fitted and provision was made (in the Bf 109E-0) for four MG 17 machine guns, two in the fuselage troughs and two in the wings, firing outside the propeller arc. Early anticipation of a future fighter bomber role led to the incorporation in the Bf 109E-1 of the necessary wiring and release mechanism for a range of bombs up to 250 kg (550 lb) which were carried on a ventral rack below the fuselage centre-section.

The Bf 109E-1 entered production at the end of 1938 and was widely issued to the *Jagdgruppen* based throughout Germany, after an initial batch of 40 had been shipped to Spain for use by the Condor Legion. The first of these is believed to have arrived there in late January 1939; one of the first examples to be lost in action by *Jagdgruppe* 88 occurred on 6 February when an E-1 was shot down, killing the pilot, *Uffz* Heinrich Windermuth. The Spanish batch was completed in March 1939.

To give an idea of the rapidly expanding rate at which the Bf 109 was produced is a quoted figure of 1,540 completed by the end of 1939. That same year E-1 production was complemented by

A series of pleasing in-flight views were taken of Bf 109E-3 CE-BM, an example that served Messerschmitt until 1942. (*via Robertson*)

First recipient of the Bf 109E-3 in quantity was the Condor Legion's fighter unit. This battered photo of the second aircraft of this model delivered survived the intervening years. (*via R L Ward*)

the E-2 and E-3 and once again, different sub-types were built simultaneously until such a time as contracts for the E-1 were completed. Relatively few examples of the interim Bf 109E-2 appear to have been built although these are understood to have been virtually identical to the E-3, with similar wing cannon armament. Produced primarily as a fighter, the E-3's firepower was later increased by a pair of MG-FF wing cannon to replace the lighter MG 17 weapons. There was no initial provision for an external bomb rack, although this was introduced on the E-3/B which could carry an ETC 50 or 500 bomb rack.

Exports

The late 1930s world-wide drive to modernise armed forces led numerous foreign customers to examine the products of the burgeoning German

Alarmstart activity around a Bf 109E-1 of 2./*JG* 20, which was later absorbed by *JG* 2.

One of many 'action' photos taken just prior to hostilities in 1939 and designed to emphasise the *Luftwaffe*'s high state of readiness, this view shows Bf 109E-3s of *JG* 20 on exercise.

Swiss defence units eventually flew all the main production variants of the Bf 109, the *Emil* being particularly popular. This is the first E-1 to arrive. *(H Holmes)*

aircraft industry, particularly after the publicity surrounding the Bf 109's success at Zurich. They included Switzerland, Yugoslavia, Hungary, Slovakia and Rumania, all of which recognised the outstanding quality of the Bf 109E, a fact not exactly suppressed by German propaganda.

So keen was the Swiss government to have Messerschmitts replace its outdated Dornier fighters that it became, in late 1938, the only country outside Germany to purchase the Bf 109D. Ten examples of the D-1 were delivered pending the supply of 80 E-1s and E-3s intended to fully update the fighter units of the *Fliegertruppe*. These latter, which eventually equipped six *Fliegercompagniene*, also saw limited air action against both Axis and Allied aircraft which strayed, for various reasons, into Swiss airspace. The Swiss Bf 109Ds and Es were supplemented by a small number of ex-*Luftwaffe* Bf 109F-4/Z variants in 1944, the German fighter successfully defending the country's neutrality throughout the war years and not being finally retired until 1948/49. Some wheeling and dealing subsequently led to the wily Swiss acquiring supplies of Bf 109Gs, about which, more later.

Yugoslavia's association with the Bf 109 began in January 1938. Negotiations nearly foundered but a deal was finally struck in April 1939, the Yugoslavian premier offering Germany vital iron ore, chrome and copper as payment for 50 Bf 109E-3s and 25 spare DB 601 engines. A second contract covered another 50 *Emils* but in the event, the Yugoslavian air arm, the *Jugoslavensko kraljevsko ratno vazduhoplovstvo* (*JRKV*) received only 73 aircraft and German 'after sales' support was apparently so poor that many of these remained grounded, awaiting spares that never materialised.

There had actually been a marked German reluctance, due to distrust over the country's support for Hitler's policies, to sell Yugoslavia the Bf 109. *General Oberst* Hermann Göring, who handled the negotiations, made numerous attempts to convince the Yugoslavs that the Messerschmitt fighter was not a desirable option, citing its weak undercarriage and demanding handling qualities which, he believed, would make it too dangerous for *JKRV* pilots to master.

Yugoslavia persisted and the deal went through, although when the 6th Fighter Regiment finally received the first Bf 109E-3s in 1939, a spate of crashes appeared to make Göring's doubts ring true. It appears that a number of Yugoslav E-3s were returned to Messerschmitt for investigation, but exactly what the reported defects were, remains a minor mystery. To fly in Germany the aircraft had to have civil registrations and these letters were simply painted over the *JKRV* letter-number codes on the fuselage sides. One example, with the registration D-IWKU applied over the Yugoslavian code 'J-55', was written off in a crash in 1940.

After the surprise announcement that a German-Soviet Pact had been signed in August 1939, five Bf 109Es were sent to the Soviet Union. It is presumed that these were intended for evaluation only and that the Soviet Air Force drew up a comprehensive set of performance figures some two years before such information would take on a new importance.

Japan received three Bf 109E-3s following a demonstration by test pilot Willi Stor for a Japanese military mission at Regensburg in 1941. Stor went to Japan in May that year, ostensibly to supervise production of the German fighter by the Kawasaki works and to train Japanese pilots. The trio of E-3s duly arrived by sea and were assembled and test flown from Kawasaki's works airfield at Gifu. But by the time of Pearl Harbor, production plans had been dropped. Stor remained in Japan where he subsequently became involved with licence-production arrangements for the Me 410. This plan was also stillborn although the fact that the Bf 109Es had been observed on test flights led to a mistaken belief that the Bf 109 had entered JAAF service.

In similar fashion to the RAF reporting the Heinkel He 113 in *Luftwaffe* service in Europe, so American pilots were subsequently to report 'seeing' the German fighter in the Pacific. To be on the safe side, Allied intelligence allocated the wartime Pacific area codename 'Mike' to the Bf 109, even though no Bf 109s saw Japanese operational service.

Chapter 2
First Blood

Of all the powers that supplied aircraft and/or pilots to fight in Spain during the civil war, few were to learn as much as did Germany. Hitler's support for Franco's Nationalists was to bring him a number of benefits, including vital raw materials supplies and an agreement on the use of bases for Germany's U-boat fleet. It was initially something of a gamble that Franco could win the war, which was why German military backing grew in scope as the conflict widened.

It was also realised that although the *Luftwaffe* had devised a number of tactics for use in modern war, the theories and the machines that would execute them had never been put to the test. Hitler seized the opportunity offered by Spain's self-inflicted strife to send as many fighter pilots and bomber aircrew, plus ground personnel, as were required while the conflict lasted. These men would serve in Spain on a rotational basis as members of the grandly-named Condor Legion. The Legion came to be considered as an elite force and it did indeed become something of an 'air force within an air force', staffed almost entirely by German nationals. Elements of the Legion were to remain in Spain for 29 months and at its peak, front line strength exceeded 100 aircraft.

The Condor Legion's fighter force, *Jagdgruppe* 88 (J 88), began combat operations with He 51 biplanes, subsequently received examples of the Bf 109B, C and D and was being re-equipped with the E model by the end of the conflict. J 88 comprised four *Staffeln* (Squadrons) known as 1./J 88, 2./J 88 etc. The Germans realised that along with reliable aircraft, a modern air arm needs well trained personnel, a firm operational infrastructure and good leadership – attributes which were not exactly over-emphasised by their Spanish allies!

In Spain the *Luftwaffe*'s expeditionary force set out to put into practice the professional dedication which had been stressed throughout pilot training and to send home as much valuable data as possible on the strengths and weaknesses of its own aircraft in all the important categories. More important perhaps, it was able to evaluate many aircraft flown by the Republican side, particularly those supplied by the Russians, and forward regular reports to Berlin.

While they were not alone in learning lessons in the contested Iberian skies, the Germans arguably put their results to the most practical use. They proved the usefulness of the dive bomber and the effectiveness of ground strafing, the importance of medium bombers maintaining a tight formation to achieve a concentrated bomb pattern when deploying against specific targets – and the superiority of modern monoplane fighters compared to biplanes.

When the war began on 18 July 1936 and Republican and Nationalist battle lines were drawn, the outside world responded to either left- or right-wing causes. The Republican government continued to elicit the loyalty of the rank and file of the *Aeronautica Militar Espanola*, although almost the entire officer corps defected to Franco. The government air arm enjoyed a numerical superiority in aircraft although most Spanish warplanes were obsolescent. Appeals to foreign powers to supply more modern aircraft were met, mostly by more obsolete types, as relatively few monoplanes then existed in Western Europe. Numbers were however, deemed important, as both Republicans and Nationalists received war material and men. By the autumn of 1936, a steady flow of aircraft from abroad was underway.

At the outset the Nationalists counted their support from the population of northern and western areas of Spain plus Cadiz, Seville and Cordoba in the south, towns located in a narrow belt to the frontier with Gibraltar. In addition, the islands of Majorca and Ibiza and the Moroccan colonial territories were sympathetic to Franco. The Republic held the Biscay coastal strip around Bilbao and Santander, the entire east coast and most of central and south-eastern Spain. Government forces could also count on support from Basque nationalist areas on the northern coast.

Numerous examples of French, Italian and Russian aircraft types were deployed in the fighting for Madrid, the Soviet Polikarpov I-15 and particularly, I-16 fighters, tipping the balance in the Republicans' favour. With his outmoded German, French and Italian aircraft failing to maintain the necessary aerial superiority, Franco appealed for more modern equipment. Germany recognised Franco's government on 18 November 1936.

With the Condor Legion largely established in Spain by late November and the fighting for Madrid at a slowdown, its chief of staff, *Oberst* (Colonel) Wolfram von Richthofen, took steps to introduce the Bf 109 to the war theatre. He requested examples for operational testing and in response the V3, V4, V5 and V6 were despatched.

Various pilots including Hannes Trautloft, evaluated these prototypes against examples of the main Republican types, the results being highly encouraging, for the Bf 109 potentially outperformed them all. Despite some allowances having to be made for minor teething troubles that grounded the prototypes for differing periods, there was no doubting that Messerschmitt had produced a winner.

On 14 January 1937, Trautloft ferried the V4 to the Madrid front for further evaluation. Both the V5 and V6 arrived in Spain in January, these being assembled at Seville and made available for evaluation by J 88 personnel pending the arrival of the first three Bf 109B-1s on 14 March.

It was decided initially to re-equip a single *Staffel* of J 88 with the Bf 109B-1 despite the restrictions imposed by the fixed-pitch airscrew fitted to these early production machines. It fell to Gunther Lutzow's 2. *Staffel* (2./J 88) to introduce the Bf 109 to combat, there being seven Bf 109B-1s on hand by the time the unit moved to Vitoria. Fittingly, the first victory for the new fighter was scored by Lutzow himself on 7 April 1937.

But it was clear that even the Bf 109B-1 did not enjoy a clear ascendency over the I-15 and I-16; its armament was too light and a fixed-pitch airscrew did not provide the Jumo engine with the speed and power range increasingly demanded by modern air combat. Few fighters extant could match the I-16 for manoeuvrability as no other designer had been able to emulate the incredibly small airframe dimensions that Polikarpov had achieved.

Jagdgruppe 88 flew the majority of its early war sorties with the He 51 which by 1937, was seen to be dangerously outmoded. Pilots of the other *Staffeln* were encouraged to fly the Bf 109 whenever time permitted, for monoplane fighters had begun to feature many new items of equipment and operating procedures that were entirely outside the experience of most personnel. When early examples of the B-2 arrived, pilots also found an extra demand placed upon them for as well as its enclosed cockpit, retractable landing gear and other innovations compared to biplanes, the machine was fitted with a FuG 7 radio telephone. This item of equipment, viewed as a dispensable luxury by some pilots, was nevertheless found to be invaluable on operations, particularly when new tactical formations were being devised.

German pilots, already using the *Schwarm* (four aircraft section) and *Rotte* (pair) (both of which probably originated at the end of WWI) as the most ideal small tactical air unit, appreciated the advantages their own tactics had over the 'V' formations still favoured by other nations. German tactics were tailored closely to the speeds of monoplane fighters and the generally rapid development of air operations that Spain came to represent.

Teblada near Seville was the main engineering base of the Condor Legion and all assembly of crated Bf 109s arriving by sea was carried out there. Under 'Franzl' Lutzow, 2./J 88 continued the working-up process and by June when the entire *Gruppe* came together at Herrera de Pisuerga, a sufficient number of pilots had completed the brief 'three-phase' conversion school – two training flights followed by a third which was an operational sortie.

When the Bf 109 arrived in March, *Jagdgruppe* 88 underwent some changes to its operational *Staffeln*. The Bf 109Bs of 2.*Staffel* now sported the distinctive *Zylinder Hut* (Top Hat) insignia previously worn by the Heinkel 51s of 4. *Staffel*, which was disbanded. In the meantime 1. and 3. *Staffeln* retained the He 51.

The influx of new aircraft for the German bomber and dive bomber units as well as monoplane fighters (the Bf 109 being joined by the He 112) was a welcome boost in capability for the Nationalists, whose fighter squadrons were otherwise largely equipped with Italian biplanes. On 6 July a Republican offensive west of Madrid caught Franco off-guard. Government forces advanced rapidly, captured Brunete and threatened to cut off the entire Nationalist front sector from Madrid to Toledo. Had the enemy succeeded in isolating the main supply route from Talavera to the capital, it would have been a disaster.

At midnight on 8 July, Lutzow was informed of the deteriorating situation and ordered to move to Avila, west of Madrid, 220 miles distant. The exodus soon gathered momentum and soon all of J 88 was on the move. With their arrival within range of the front lines, the two He 51 *Staffeln* were committed to ground attacks on advancing Republican troops and anti-aircraft batteries. The Bf 109s were tasked with escort to bombers and transports.

Legion sorties came up against considerable opposition. Airfields around Madrid were crowded with I-15s and I-16s whose task was to maintain air superiority for the duration of the current offensive and the Republicans had brought up AA artillery in some strength. Heavy ground fighting in a salient at Brunete soon found the two sides engaged in a series of large-scale air battles; the Bf 109s took off three and four times a day to cover bomber sorties and reconnaissance flights. To ward off attacks on its own airfields, the Legion's fighters flew standing patrols over Madrid and Brunete to watch for Republican fighters taking off from bases situated on the other side of the capital. Readiness lasted from 0415 to 2130 hours.

At altitudes favoured by their own medium bombers, the German fighters were invariably busy as the tiny Republican interceptors strove to break up the formations. It was by no means an easy task for the Legion fighter pilots. Only if they could force the incredibly manoeuvrable *Chatos* and *Ratas* to engage above 10,000 feet did they have much chance of shooting them down for although the Bf 109B was faster than both the I-15 and I-16, it could rarely stay with either of them in a low altitude 'turning match'.

But the Germans were often able to unnerve Republican pilots with their sounder tactics, particularly when they positioned a fighter section well above the predominantly low level combats. These Bf 109s were ideally positioned to pick off stragglers from the Republican formations and when they initiated high speed dives, the enemy had little or no defence against them – indeed it was said that some *Chato* and *Rata* units actually landed before realising they were short of one or two 'tail-end Charlies'.

The German air discipline developed to prevent such an occurrence happening to their own; the superb 'finger four', with spacing based on the wingspan of the Bf 109, enabled an all-round lookout to be maintained at all times, for fighters to avoid getting in each other's way and for turns to be made safely without creating blind spots for enemy interceptors to exploit. Again the pilots of this war had taken heed of the developments at the end of the last one and adapted them accordingly.

Among the bombers heavily deployed by the Republicans at this time was the Tupolev SB-2, a type that began to appear in the J 88 victory lists with increasing frequency. Often mistakenly referred to as a Martin type, the SB-2 *Katiuska* proved a relatively easy victim for the Germans. Two were claimed on 8 July by *Leutnant* (Lt) Rolf Pringel and *Unteroffizier* (*Uffz*) Guido Honess, the latter pilot having the dubious distinction of being himself being shot down by an SB-2's defensive fire on 12 July – the first combat loss publically admitted by the Legion.

A further Bf 109 loss on the 20th was that flown by *Uffz* Dieter Hrabach. Obliged to take to his parachute, Hrabach was lucky enough to be blown across Nationalist lines and was back at his base, Avila, the same evening.

The number of German pilots scoring aerial

victories in Spain rose steadily, albeit with a degree of over-claiming that was all but inevitable in the confusing melee of a dogfight. Republican claims could likewise be erroneous – and at times there was deliberate fabrication, for it was widely believed that getting the better of one of the much-vaunted Bf 109s could be highly prestigious to the individual concerned! Among the Legion pilots who scored during this phase of the fighting were Rolf Pringel, *Uffz* Buhl, *Oberfeldwebel* (*Ofw*) Hilmann and *Fw* Boddem, with one enemy fighter apiece.

By 14 July the Republican advance had been halted and the Battle of Brunete, the first such engagement of the war worthy of the term, was over. Four days later, the Nationalists launched a counter-offensive and the task of 2./J88 was to intercept and break up enemy bomber formations threatening ground forces and lines of communication behind the front. To circumvent this the Republicans mounted a number of raids on Condor Legion airfields, including Avila.

One such operation was planned for 23 July but the attack went disasterously wrong. Although they were scrambled hurriedly, 2. *Staffel* pilots hurled themselves at the SB 2s and succeeded in breaking up their formations. Losing only one aircraft, to *Fw* Boddem, the Republicans nevertheless scattered, jettisoned their bombs and fled. Not even a splinter landed on Avila.

By the end of July the front had stabilised to such a degree that *Jagdgruppe* 88 could withdraw and return to Herrera. Franco's offensive against Santana was resumed on 14 August, his air forces having gained almost total air superiority. The Republicans lost particularly heavily in terms of fighters during the Northern campaign, two squadrons being almost wiped out.

But the SB 2s of *Grupo* 12 and 24 were active and on 27 August, the day after Santander fell to Franco, one was shot down by *Oblt* Harro Harder. J 88 continued to follow the advance westwards along the Asturian Biscay coast, using a number of small airstrips in order to remain in range of the fighting.

While 2. *Staffel* had been writing the first chapter of the Bf 109's long combat history, more aircraft had been delivered to Tablada and by the autumn of 1937, enough were on hand to equip a second. This honour fell to 1./J 88 and in September its He 51s were duly passed on to a new 4. *Staffel*, which was reformed as a ground attack unit. Gunther Lutzow bade farewell to his comrades in 2. *Staffel* for he was now to lead 1. *Staffel*, his place being taken by *Oberleutnant* (*Oblt*) Joachim Schlichting.

The Nationalists now had 15 fighter squadrons, three of them German and with 12 flying Fiats for a total strength of about 140 aircraft. For the first time, Franco could count on more fighters than the Republicans, although the latter still had about 120 machines.

Despite the experience passed on by its new *Staffelkapitän*, the early operations of I./J 88 were marred by losses inflicted, mostly by the I-16. During one dogfight with six I-16s and the last surviving I-15s (which had suffered heavy losses in combat), the *Staffel* lost two out of six Bf 109Bs which were intercepted as they flew escort to Legion He 111 bombers.

But overall the Nationalist victory in the northern campaign was highly significant. It gave Franco's side control of some 50 per cent of Spanish industry, including armament factories; 100,000 Republican prisoners had been taken and the Nationalists had inflicted some 130,000 casualties – about 75 per cent of government troops then available for combat. Also, it was in the nature of the Spanish conflict that a great many individuals did not remain prisoners for long but threw in their lot with the Nationalist cause.

There followed a lull in the ground fighting as Franco planned his strategy to secure Madrid and in November, the Legion occupied airfields at El Burgo de Osma and Almazan to support an offensive against Guadalajara. Having completed his tour, *Oberleutnant* Schlichting was replaced at the head of 2. *Staffel* by *Oblt* Wolfgang Schellmann and it was at this time that the Condor Legion had a new commander when *Generalmajor* Volkmann replaced Hugo Sperrle.

On 4 December the Bf 109B flown by *Fw* Polenz ran out of fuel after escorting He 111s bombing Bujaraloz airfield and was forced to put down on a road between Azaila and Escatron. Both pilot and aircraft were captured, the Bf 109 eventually being moved to France for a thorough evaluation.

Again the Republican forces stole a march on Franco by pre-empting any Nationalist moves when on 15 December, they moved against Teruel. Although the Legion and other air elements moved up – in J 88's case to Calamocha, 50 miles from the fighting – airpower failed to materially dictate the course of the ground war. The main cause was the weather, which brought extreme cold and temperatures well below freezing to curtail flying.

At the end of 1937 the Legion lost the expertise of *Oblt* Harro Harder, who returned to Germany with a very respectable 11 kills to his credit, one short of Wolfgang Schellmann's eventual 12 victories – a record that was, in turn, to be broken only by Werner Mölders.

The Decisive Year

January 1938 was a little over a week old when SB 2s of *Grupo* 24 – the aircraft of *Grupo* 12 having been absorbed due to heavy attrition – attacked Calatayud and La Gea. On the 12th, another SB 2 was shot down by Bf 109s over La Cenia airfield. Two were actually claimed, both by 2./J 88. On 18 January, Schellmann shot down an I-16, which helped the Legion achieve the milestone of 100 kills since the conflict began.

A vicious running battle over Formiche on 7 February resulted in *Grupo* 24's SB 2s being decimated by 1. and 2. *Staffeln* of J 88 which shot down four within minutes. While other German pilots claimed kills and were duly credited, subsequent analysis showed that the victor over all four *Katiuskas* had been one pilot, *Oblt* Wilhelm Balthasar. This feat, rarely achieved in Spain, brought Balthasar's total to seven, a score he would not improve upon before returning home.

Due to the opposing sides flying exactly the same or very similar types, aircraft recognition was a challenge throughout the war and an incident on 21 February highlighted the problems. Nationalist Fiat CR 32s came upon a mêlée of *Chatos* and *Ratas* dogfighting with Bf 109s and waded in. Much to the chagrin of the Italians, the German pilots apparently failed to recognise aircraft in Nationalist markings and promptly manoeuvred to deal with this new 'threat'.

Neither side suffered but the incident was one of a number that did nothing to improve already poor relations between Germans and Italians.

At the end of the Teruel campaign the Bf 109s had destroyed 30 enemy aircraft, an impressive record for 'Franco's fire brigade' as the Legion came to be known. The town of Teruel was in Nationalist hands by February and Franco prepared for a new spring offensive.

Aragon

Starting 9 March, Franco made a two-pronged attack north and south of the River Ebro. In eight days, 2,700 square miles of territory had been taken and several air bases, including La Cenia, had been captured. The Legion lost a Bf 109 on 10 March during a series of Republican attacks on aerodromes at Bujaralos, Candasnos and Caspe.

Another *Jagdgruppe* 88 base move was made on 27 March, the Bf 109 *Staffeln* occupying Escatron. From this airfield sorties were launched against Lerida, where several I-16s were destroyed on the ground. Such strafing attacks by the well-armed Bf 109s not only proved highly profitable in containing Republican aircraft but the generally modest enemy defences did not expose the fast German machines to undue risk. Only one Bf 109 was lost in ground attacks up to 1 April.

Frequent base moves however caused some drop in Legion aircraft serviceability, a factor that in March 1938 was compounded by an event outside Spain. The Austrian *Anschlüss*, a calculated risk by Hitler, placed the German armed forces on full alert. The move to bring Austria under the umbrella of the *Reich* could have precipitated hostile moves against Germany and with relatively weak forces at home, a temporary embargo was placed on armament shipments. This included replacement aircraft for the Condor Legion in Spain, although it did not last.

On 15 April the Nationalist offensive finally succeeded in cutting Spain in two and La Cenia, in the Saragossa region of the Ebro valley, became the main base of the Condor Legion. The focus of the ground fighting was now south, along the Mediterranean coast where a continuing call for air support created some difficulties. Republican

20-mm AA guns did not make the task any easier – in fact Nationalist losses were heavy enough for ground attack sorties to be suspended if they could not count on a sizeable fighter escort. Attrition had robbed *Jagdgruppe* 88 of almost half its Bf 109s which were reduced from a combined total of 30 machines in the two *Staffeln* during the winter months, to 16. The Legion's opponents on the Republican side included a number of non-Spanish nationals and the pilot of an I-16 shot down by *Hauptmann (Hptm)* Gothard Handrick on 18 May was found to be an American.

The Bf 109s were increasingly engaged on bomber interceptions and on 2 June, 2. *Staffel* claimed five SB 2s from a force that attempted to surprise La Cenia. Only two bombers were actually destroyed by the Messerschmitts, with two others falling to flak. On the 10th, Lt Hans-Karl Meyer of 1. *Staffel* shot down another SB 2 over Lucena del Cid. The early part of June was marked by a series of combats, most of which involved the Bf 109s. Republican losses for the 10th included two *Ratas* as well as Meyer's SB 2 and three days later, the Germans claimed two more *Ratas* and five *Chatos* over Castellon.

On the 14th a Bf 109B flown by *Lt* Henz was involved in an unusual incident. His aircraft hit and trailing coolant, Henz made a landing on the right bank of the River Miares and took cover. Circling Republican fighters resisted the urge to strafe the German aircraft, as only one other intact example had yet fallen into their hands and which, as related earlier, had been quickly despatched to France.

As the most modern fighter in Spain the Bf 109 was highly prized and the Republicans clearly intended to capture it intact, perhaps to make a more thorough evaluation than had previously been possible. A number of Republican fighters landed nearby and their pilots were intent on examining it when six Bf 109s came in low over the river. Turning to starboard, they circled the grounded fighter and systematically strafed it. It was on fire within minutes.

It was also in June that the Condor Legion underwent another minor crisis. With enough Bf 109s to re-equip 3./J 88, a new *Staffelkapitän* had to be found to replace Adolf Galland, who was due to return home after a tour on the He 51. A pilot familiar with combat conditions in Spain was naturally preferred, but of the two candidates selected, one proved unsuitable and the other was promptly killed in a flying accident. There was nothing for it but to choose someone new to the war theatre. His name was Werner Mölders.

Mölders' reformed 3. *Staffel* would fly the Bf 109C, five examples of which had arrived in April. As it transpired, these were the only Bf 109Cs to serve in Spain and with their armament increased to four machine guns, they were considered well suited to combat the so-called 'Super Mosca' version of the I-16, which also had four guns.

Both Bf 109 *Staffeln* joined forces to bring down seven Republican fighters on 13 June, five I-15s and two I-16s falling to their guns over Castellon de la Plana, which was captured by the Nationalists two days later. These victories more than made up for the loss of *Lt* Henz and his aircraft, plus that of *Lt* Plieber, who was wounded when his machine was written off in a forced landing.

Meanwhile, Mölders threw himself enthusiastically into his new command and on 15 July in company with Wolfgang Lippert and Walter Oesau, he opened his own account when the first kills by 3. *Staffel* were made with the Bf 109. Mölders barely had the target I-16 in his sights before opening fire. He missed and dropped back, cursing his ineptitude and vowing not to repeat the mistake. Selecting another *Rata*, he closed in to point-blank range. Only when the tubby Republican fighter filled his windscreen did Molders press the gun button. This time, the results were spectacular. The I-16 shook under the impact of machine gun fire and flicked over and down, trailing smoke. Mölders watched it hit the ground. Another I-16, Mölder's second kill, fell on 17 June and in fact all his victims were to be enemy fighters.

Five *Ratas* and five *Chatos* were shot down over Segorbe by Bf 109s on the 18th, the Germans flying in company with Italian-manned CR 32s. The day's score was not entirely one-sided however, for three CR 32s were lost. For the day's total, the Republicans claimed two Bf 109s and eight more CR 32s, some of the victims apparently falling to a force of 25 I-16s which saw action later that day.

Mölders' rapid scoring rate continued on 19 June when he shot down his third Republican fighter. Fellow pilots Ebbinghausen, Hein and Tietzen also scored one apiece from a formation of 18 SB 2s with three *Rata* squadrons escorting. In combat over Sagunto, no losses of Legion aircraft were reported.

In such engagements, Mölders would occasionally relinquish the lead position in his *Schwarm* and fly on the flank. This passed responsibility to other pilots and at the same time, gave Mölders maximum flexibility to break away if a likely target presented itself. A man gifted with remarkable eyesight, he could often spot other aircraft before anyone else in his formation.

Mölders readily appreciated the advantages he 'finger four' formation presented for high speed fighters and adopted and honed it with great enthusiasm. Although he is often credited with 'inventing' the formation, it was in fact in vogue before he arrived in Spain and had, as mentioned earlier, almost certainly originated before the war.

On 17 July the entire Republican SB 2 *Grupo* attacked Sagunto, the Republican force having been swelled by further aircraft deliveries from Russia. Intercepted by 3./J 88, one SB 2 was shot down – although no Legion claim was filed. Schellmann scored his seventh and eighth victories on the 20th, both of these being I-16s. The 23rd saw all three Bf 109 *Staffeln* putting all available machines into the air to repel a large force of Republican fighters reported over Viva. From the estimated 40 *Chatos* and *Ratas*, the Legion despatched six without loss.

On 25 July the Republican offensive across the Ebro, aimed at drawing off Nationalist pressure on Valencia, sparked some of the bloodiest fighting of the war. It was to denigrate into a slogging match of attrition which included six Nationalist counter-attacks on the bridgehead south of the river being repulsed. Both sides deployed bombers to force a conclusion, the Legion to cut bridge crossings and the Republicans to attack ground forces and enemy aircraft on their airfields. By attempting the latter, the government forces often lost heavily, as *Grupo* 24 experienced on 27 July while attacking Vinaroz, Valderrobles and Maella. Falling foul of Legion Bf 109s, the bomber force lost three of its number, although only one kill was claimed, by Walter Oesau.

More strength was added to *Jadgruppe* 88 in August with the arrival of five Bf 109Ds at La Cenia. New equipment was more than welcome but this new version appeared to be a mixed blessing. While it had the four-gun armament of the Bf 109C, the powerplant had reverted to a carburettor system for the Jumo 210 engine. Something of a hybrid, the Bf 109D was nevertheless to be the most numerous model sent to Spain to actually see combat, the Legion taking delivery of 39 examples. The doubts proved groundless, for the D model's Jumo 210D (similar to that fitted in the Bf 109B) delivered sufficient power at the low to medium altitudes at which air combat usually took place.

On 12 August, three more SB 2s were credited to 1./J 88, the victors being *Uffz* Helmut Brucks, *Lt* Otto Bertram and *Hptm* Wolfgang Schellmann. Yet another fell to Mölders on the 23rd, a day on which four separate air battles were fought by the Bf 109s, a clear indication of the intensification of operations.

On 30 August Mölders took 3./*Staffel* to Pomar and for about three weeks there was a lull in activity. Not until 13 September did the Legion's Messerschmitts again engage in combat, but four *Ratas* were shot down that day. Another period of inactivity followed and the Legion took the opportunity to again reorganise its personnel. Handrick relinquished command of *Jagdgruppe* 88 to *Hptm* Walter Grabmann, the three *Staffeln* now being commanded by *Oblts* Reenst (who had succeeded Schellmann), Krok and Mölders. Walter Oesau was the *Gruppe* Adjutant.

On 4 October a Republican ground attack on La Cenia managed to destroy several fighters. Although Mölders shot down a *Rata*, Otto Bertram went down. It was at this time that 3. *Staffel* introduced the *Vierfingerschwarm* or 'finger four' as a standard – rather than experimental – fighter formation, one that was to pay off handsomely not only in this conflict, but in the future. On 15 October, Mölders scored a *Rata* double to bring his score to eleven.

There was yet another lull in operations at La Cenia when on 18 October, heavy rains brought flooding. The last day of the month saw Mölders bring down another two Republican fighters for his twelfth and thirteenth victories of the war; his time for returning to Germany fast approaching, Mölders capped his string of victories with one more on 3 November.

November recorded further organisational changes, with Wolfram von Richthofen taking command of the Condor Legion as General Volkmann's service period in Spain was all but complete. Thus Richthofen was able to see at close quarters how the Legion had developed during its sojourn and most of what he saw convinced him that if the *Luftwaffe* adopted the tactics that had been proven in action, it could conduct a series of decisive tactical campaigns in concert with ground forces.

That the fighters of the Legion were a painful thorn in the side of the Republicans was evidenced on 16 December when yet another attack on La Cenia was attempted. But the Bf 109s were airborne in time to intercept – due it was believed to the work of spies forewarning the Germans of enemy intentions – and two were shot down by *Uffz* Herbert Schob and *Lt* Helmut-Felix Bolzof of 2. and 3. *Staffel* respectively.

The second Ebro offensive, which did not end until 16 November, all but decided the outcome of the war. This was certainly the view in Berlin, where the requests to re-equip the Legion fighter squadrons with the latest Bf 109E model were coupled with the need to modernise the remaining Nationalist fighter units, the bulk of which were still flying the CR 32. To meet this need, the German government reiterated the terms of the previously agreed deal whereby the full cost of keeping the Legion in Spain was borne by the Nationalist cause and that payments be made in iron ore. A grateful Franco agreed entirely with German terms.

By 23 December, the date of the opening of the last Nationalist offensive of the war, Franco's advance on Barcelona could continue to rely on the material assistance of the Condor Legion. *Jagdgruppe* 88 had on the opening day of the offensive 37 of the total 55 Bf 109s already despatched to Spain. Most of these were decidedly 'war weary' but the arrival of the first examples of the Bf 109E a few days after the battle started, prevented any appreciable reduction in strength.

The decision to send examples of the most modern fighter Germany possessed was not entirely understood at home, where some *Luftwaffe* units were still flying the older model Bf 109s or obsolescent biplanes. But on the strength of the iron ore deal, Iberian interests took immediate precedence.

Among the final victories scored by Legion pilots in 1938 were three more SB 2s shot down on 28 December by *Lts* Wilhelm Ensslen, Heinz Bretuntz and Karl-Wolfgang Redlich. On 4 January 1939, an SB 2 became the second of Redlich's kills and the last of this type to be destroyed by the German fighter force. Four days later, the second phase of the Catalonian offensive began and on the 12th the Legion fighters attacked enemy aerodromes, among them Vills, El Vendrell and Vilafranca del Panades.

With the fall of Tarragona on 15 January, the road to Barcelona was wide open and the city was in Franco's hands by 26 January. On the 21st the Bf 109s had occupied Vills but in a few days they were on the move again, to Saball, the main task at this time being to prevent Republican aircraft escaping to the South.

Air action became increasingly scarce in the last few weeks of the conflict but on 31 January the Legion's last combat loss was recorded when a Bf 109 was shot down over the Spanish frontier. Flying sorties that were in the nature of mopping-up operations, the German fighters strafed retreating Republican forces. On 5 February the Bf 109s, in company with Italian Fiats, made a devastating strafing attack on Figuras aerodrome, leaving both installations and aircraft in flames. The attack took less than two minutes. On the 6th, *Uffz* Windermuth's Bf 109E-1 was shot down.

Franco's victory was now only days away. Air combat had dwindled but on 5 March *Oblt* Hubertus von Bonin, Mölders' successor as *Staffelkapitan* of 3./J 88, claimed the Condor Legion's last kill, an 1-15 which fell over Alicante for his fourth personal victory.

Late March saw the first unit of the Spanish Air

Force re-equipping with the Bf 109B, as sufficient supplies of *Emils* had been received for the Legion. The German fighters had by then moved base yet again, to Bercience (Torrijos). It was from this base that *Jagdgruppe* 88 flew its last sorties of the Spanish Civil War on the morning of 27 March 1939. Three days later, the ceasefire was announced.

Their duty in Spain completed, the majority of Condor Legion personnel were packing to leave for home by April. In Germany news of their exploits brought forth medals and promotions and the sun-bronzed veterans – quickly nicknamed 'Spaniards' – were feted at parades, receptions and galas. Their part in the first war in which a fascist regime had triumphed by force of arms had been significant, the Legion's final victory tally being 340 enemy aircraft destroyed in air combat.

While Nationalist aerial victories had been scored by other fighters, the Bf 109 had emerged as the most successful German type. As much if not more important for the future deployment of the *Luftwaffe* was the invaluable experience gained by dozens of pilots, most of whom were now in a position to share their knowledge and hone the training of fighter, bomber and ground attack pilots. It would not be long before this knowledge would be put to a far more challenging test.

When the Condor Legion left Spain in 1939, its remaining Bf 109s, both Jumo and Daimler-Benz engined examples, were passed to the Nationalist air force. These machines were to help guard Spain's neutrality during the turbulent years of World War II – but the country's more peaceful association with the Bf 109 story would extend well beyond 1945.

Chapter 3
East–West Triumph

It was not until December 1938 that *JG 234* *'Schlageter'* based at Cologne became the first *Luftwaffe Jagdgeschwader* to receive the Bf 109E-1 and E-3 but in the following ten months or so up to September 1939, a total of 14 *Gruppen* and two *Staffeln* had been similarly re-equipped, a further *Gruppe* and a *Staffel* then being in the process of conversion. Six *Zerstörer Gruppen* retained the Bf 109D-1, pending the delivery of the Bf 110, the type that had equipped only three *Gruppen* by the outbreak of war. Thus the number of Messerschmitt fighters in *Luftwaffe* service had leapt from 171 in September 1938 to 1,059 one year later.

Displaying not unusual replacement panels after doubtless hard flying over France, this Bf 109E-3 of *JG 2* shares a dispersal with an Hs 126, the *Wehrmacht*'s 'eyes in the sky.'

One of JG 52's *Emils*, complete with rear-view mirror. Some aircraft were fitted with these small but vital aids to survival during the battles over France and England in 1940. (*H Holmes*)

With Hitler's impending military action against Poland, the bulk of the Bf 109 force was retained in Germany in anticipation of hostile action by nations which had given assurances to the Poles to take up arms should their country be attacked. Of the above total of 1,059 Bf 109s (1,015 of which were serviceable in late August), 459 were based on airfields in the Eastern areas of Germany while 600 were available in the West.

With the Bf 109E largely reserved for home defence, the two *Luftflotten* (Air Fleets) which had responsibility for operations over Poland drew almost entirely on the strength of the *Zerstörer Gruppen* equipped with the Bf 109D. These were: I./ZG 2 (operating under the temporary designation JGr 102); I. and II./ZG 26 (no re-designation); III./ZG 26 (JGr 126); I./ZG 52 (JGr 152) and II./ZG 76 (JGr 176). To complete this list, II./ZG 1 '*Wespen*', equipped with the Bf 109E, was known as JGr 101. I *Gruppe Lehrgeschwader* 2, equipped with the Bf 109E, was an operational test unit which operated independently from the main *Luftwaffe* formations but which was attached to one of them on a rotational basis, in the case of Polish operations, to JG 77. The unit subsequently flew a variety of first line aircraft in order to explore their strengths and weaknesses under combat conditions.

In the early hours of 1 September 1939, *Luftwaffe* pilots were put on 15-minutes readiness for war operations against Poland to begin at first light. From the start of the attack, dive and medium bombers, destroyers and fighters ably

Trio of Bf 109E-4s on a ferry flight complete with long range fuel tanks and full *Stammkennzeichen* (factory identification codes) in the 'VL' series. The second aircraft is VL-AL. (*Holmes*)

supported the *Panzers* which knifed into the Polish armed forces in a ruthlessly efficient demonstration of a new form of warfare – *Blitzkrieg* (Lightning war). Having proven the feasibility of their close air support tactics to a great degree in Spain, German aircrews generally found Poland a markedly different theatre of war. For one thing, pilots who had flown for Franco's Nationalists were ill-prepared for the ferocity of the reaction by the Polish fighter squadrons. Pilots of the diminutive PZL 11s flung themselves on the invaders, some individuals not shying away from actually ramming their targets in order to bring them down – rarely had such actions been observed in Spain.

Enthusiasm for the *Zerstörer*, exemplified by Messerschmitt's 110, led the high command and particularly Göring, to task this one type with much of the escort, ground strafing and responsibility for gaining air superiority over the Polish Air Force. In the event, the heavy cannon armament of the Bf 110 was used to good effect, the weight of fire proving quite devastating against Polish targets on occasion.

With the *Zerstörerflieger* having the situation well in hand, relatively few single-seater fighter pilots, even those who had scored victories in Spain, were in a position to add to their scores, certainly not in great numbers. And of course, the duration of the Polish onslaught was so brief that many pilots had little or no chance to meet the enemy in combat.

Although the *Luftwaffe* directly deployed 1,581 aircraft against Poland, only 210 of these were Bf 109 and Bf 110 fighters. The bulk of the rest, including Ju 52s, were bomb-carrying machines. *Luftflotte* 4's I./ZG 2 did particularly well with its Bf 109D-1, the unit's base at Gross-Stein near Oppein in Upper Silesia being the jumping-off point for the first war sorties. *Hptm* Hannes Gentzen led I *Gruppe* and was credited with seven victories.

Gentzen, a renowned pre-war sport flyer, later wrote about his experiences over Poland. He stressed the effectiveness of the green and brown camouflage applied to Polish fighters (a subject that would be studied at some length by the *Luftwaffe*) and described a combat on that first day of the war. Flying near Lodz at 3,000 feet, Gentzen's Messerschmitts came upon two PAF fighters, one higher than the German formation. Part of his *Staffel* broke downwards and Gentzen attacked one of the enemy fighters. His fire appeared to have hit the machine's engine and it fell away.

The Bf 109s followed the aircraft down and the Polish pilot led the Germans to a hitherto unknown base, one of the many used by the PAF to disperse its fighters. According to Gentzen, this airfield was camouflaged well enough to remain hidden from air reconnaissance, had the enemy pilot not sought refuge there.

Wide dispersal of combat aircraft, a sound enough principle in itself, tended to hamper the Poles, who found that with their communications gone, co-ordinating interceptions over vital areas was difficult, if not impossible to achieve. Many contacts were however made and among the German pilots who found themselves on the receiving end during this first phase of modern air warfare was Hans Philipp. He was obliged to make a forced landing on the first day but was uninjured.

Philipp and countless other *Jagdflieger* came to realise that although the Bf 109 might have had a delicate undercarriage, prone to collapsing, the pilot's cockpit area was strong enough to withstand considerable damage in a crash. Providing

'Black 16', a DB-601-powered *Versuchs* Bf 109 passed to a fighter training school, shows off its patched up undersides. The tail band and spinner were yellow.

that the pilot was well strapped in (a pre-requisite for survival in crash landing most aircraft) he could expect to extricate himself from the wreckage providing there was no fire.

The Bf 109's weakest point was just aft of the cockpit and in a heavy landing the fuselage would invariably shatter at that point, leaving the heavier centre section containing the cockpit, to lose momentum quickly. The light fuselage construction did result in some spectacular 'spreading it around the field' crashes, with the fuselage, cockpit, engine and wings occasionally ending up in different corners – but the cockpit area remained intact more often than not, a plus factor highly appreciated by the men who flew the 109 in combat.

In six weeks of fighting, the Germans destroyed 260 enemy aircraft in combat over Poland, this out of total lost to the PAF of 330, the balance being aircraft damaged beyond repair and/or abandoned on captured airfields. *Luftwaffe* losses were 285 aircraft destroyed, with 279 severely damaged. Aerial victories by pilots flying the Bf 109D were a modest 13 compared to 94 by Bf 110 crews. Bf 109 losses were a substantial 67 to all causes, but only 12 were claimed by Polish fighters as definite kills, with another seven falling to bomber gunners. At least some of the losses could be attributed to the inexperience of the majority of the German fighter pilots, the sometimes unforgiving nature of the Bf 109 when deployed from unfamiliar airfields and bad weather, which can have a disruptive effect on air operations, however well planned.

Against the RAF

With Britain and France at war with Germany from 3 September, the replacement of *Luftwaffe* aircraft lost over Poland took a high priority. Yet there was hardly an air of panic on the 'home front', which had by the close of the Polish campaign, seen no direct military action against Germany itself. Nevertheless, the *Jagdwaffe* could not expect hostile action to be delayed for long.

A familiar scene wherever the Bf 109 operated was the ground loop, due either to combat damage or the type's inherently weak main oleo legs which were prone to folding up. In this view an E-1 of III./JG 2 has come to grief.

In the event it was the RAF that opened the offensive, such as it was in 1939. As the principal *Luftwaffe* interceptor, the Bf 109 had been deployed around the country, and among the units with a special responsibility to guard the most vulnerable approaches to Germany was II./JG 77, based at Nordholz in the 'North Sea Triangle'.

With a vital need to protect her sea lanes from surface raiders and U-boats, Britain had to know the whereabouts of the capital ships of the *Kriegsmarine* at all times. Consequently, aerial reconnaissance of and attacks on units of the German Navy, were given top priority, despite the fact that the RAF had neither the right aircraft nor suitable weapons to cripple, much less sink, well-protected capital ships. Even finding the ships in port proved a difficult enough job without the German fighters and *Flak*, which were invariably active if the weather was anything above appalling.

Bad weather with good cloud cover aided the British bombers but an overcast could hide defending fighters just as easily and in the late months of 1939 JG 77's Bf 109s quickly got the measure of what was arguably the RAF's best bomber of the period, the Vickers Wellington. As early as 4 September, pilots of II./JG 77 scored their first victories. *Feldwebel* (Fw) Alfred Held and Hans Troitzsch each shot down a Wellington that morning while *Lt* Metz despatched a Blenheim later in the day.

Action on Germany's border with France was soon to prove equally satisfying for the *Jagdfliegern*. Here too air combat was initially sporadic but on 8 September Werner Mölders, leading a *Schwarm* of Bf 109Es of I./JG 53, clashed with six Curtiss Hawk 75As of the *Armée de l'Air's CG* II/4. Of the foreign aircraft ordered by France before the war, only the Curtiss fighter had been delivered in time to equip front line units and it was by the standards of the day, a capable fighter. After briefly engaging the Bf 109s, the French claimed two German fighters shot down. This was not the case and it is almost certain that the characteristic black smoke emitted by the Bf 109's Daimler Benz engine when the throttle was opened, was the cause of the error. The Frenchmen would not be the last to believe they had crippled if not destroyed a Bf 109 diving away at full throttle. The engagement was inconclusive, apart from Mölders' own machine sustaining engine damage that forced him to crash-land at Birkenfeld.

Later in the day, a *Rotte* of Bf 109Ds of II./JG 52 intercepted a French ANF Mureux reconnaissance flight and *Lt* Paul Gutbrod shot down one of them. That this *Gruppe* was still flying the *Dora* model Bf 109 reflected the steady but still incomplete re-equipment of the *Luftwaffe*, a state that would have been very familiar to most European air arms at that time; virtually all of them were having to ride out a slow trickle of new and improved machines to front-line units and most were obliged to fly at least some sorties in obsolescent aircraft. The war had overtaken numerous re-equipment programmes both implemented and planned.

Gruppen of JG 52 and 53 were active over the Western Front during this period, together with smaller units that retained the temporary designations introduced at the time of the Polish campaign. Pilots of *Jagdgruppe* 152, (alias I./ZG 52) were credited with the destruction of the first RAF day bombers over the continent on 20 September when two Fairey Battles of No. 88 Squadron were shot down west of Saarbrucken. Otherwise, the opening weeks of the war on the Western Front brought such sporadic action that the Germans jocularly nicknamed it the '*Sitzkreig*'. To the Allies it was the 'Phoney War'.

Air action on any scale during the winter of 1939 was more or less confined to Germany's North Sea coast and on 18 December a force of Wellingtons raiding the Heligoland Bight area suffered grievous loss, again at the hands of pilots of II./JG 77. The Bf 109Es were directed onto their target for the first time by a new *Freya* radar situated on the island of Wangerooge. This epic air battle, from which 12 out of 22 Wellingtons 'failed to return', was an enormous tactical victory for the *Jagdfliegern*; added to the loss of five out of 12 Wellingtons on 14 December, it led to a completely new operational plan for Bomber Command. Full scale daylight sorties were seen to be far too costly in the face of German fighter interception and from then on, the main weight of the RAF bomber offensive was to be flown at night.

The stalemated situation of the exceptionally harsh winter of 1939-40 gave the German *Jagdfliegern* ample opportunity to practice their skills on training sorties and indulge in mock dogfights, any activity over the Franco-German border invariably being in the nature of a chase. Neither side wished to precipitate action in contravention of orders – but the French had to gather intelligence on German movements and many risky reconnaissance flights were undertaken in order to obtain it.

Improved *Emils*

Having standardised on the DB 601 engine, the Bf 109E was subject to considerable development. The E-3 was the most numerous sub-type in the front line *Gruppen* by early 1940 and that May the E-4 fighter emerged with a strengthened cockpit canopy framing and windscreen. The loaded weight of the E-series had inevitably risen, from 5,667 lb for the E-1 to 5,747 lb for the E-3, an increase of some 700 lb over the C-1, which had a maximum loaded weight of 5,062 lb.

There were two sub-variants of the E-4 series apart from the standard day fighter, the E-4/B fighter bomber and the E-4/N, which had a 1,175 hp DB 601N engine. Some examples of the latter were operated as fast reconnaissance fighters carrying an external camera pod in place of a bomb or drop tank, while the Bf 109E-5 (with the DB 601Aa engine) and the E-6 (DB 601N) were intended primarily for the reconnaissance role.

Along with the conventional ventral bomb rack(s) some E-4/Bs were adapted to carry a tray which extended from the engine firewall back to fuselage frame 2 and contained cells for up to 96 SD-2 anti-personnel bombs. These, the notorious 'butterfly' bombs, had to be handled carefully as they had a reputation for being unstable with a propensity to explode without warning even when being loaded on the parent aircraft. They were intended for 'seeding' large areas, particularly airfields and were quite effective in use, as the safe disposal of those that had not detonated could be quite hazardous.

Attachment points for a 300-litre drop tank were incorporated as standard on the Bf 109E-7 which was otherwise similar to the E-4N. Additional armour was added to the E-7/U2 to increase its suitability for the ground attack role and the E-7/Z's engine had a GM-1 nitrous-oxide boost system. Engine power was further increased in the Bf 109E-8 which had the 1,300 hp DB 601E and the E-9 reconnaissance fighter carried one Rb 50/30 or two Rb 32/7 cameras.

For its time, the Bf 109E was heavily armed; it possessed an impressive altitude performance and was manoeuvrable enough; its engine power was equal to that of most of its contemporaries and the fuel injection system had been proven to give superior and reliable performance in all flight attitudes and speed regimes. If there was one major drawback with the Bf 109, it was a lack of range. In this it was also very much on a par with its contemporaries and by no means the worst in comparison with the fighters of other European countries.

It is certain that with very few exceptions, the eager, confident young pilots who had been trained to fly the Bf 109 were delighted with it – and that was a factor that no designer could build into an aircraft. Detractors might have dismissed this as a mere – even imagined – psychological advantage and an empty product of imaginative propaganda. But there is little doubt that those countries opposing Germany and potentially faced with meeting the 109 in substantial numbers soon learned that to take it for granted was extremely foolhardy if not fatal.

Early in 1940 the composition and strength of the *Jagdwaffe* was further rationalised, a move that saw almost all the pre-war fighter unit numbering eliminated. JG 2, 3, 26, 52 and 53 were all brought up to three *Gruppen* strength each with three *Staffeln* numbered consecutively through the *Geschwader* while JG 70, 71 and 72 were disbanded. A second *Gruppe* was added to JG 51 and I./JG 54 had been raised by February, the month that saw I./ZG 52 relinquish its JGr 152 designation, with JGr 101 and 152 being dropped the following month. I./JG 27 had already been formed, on 1 October 1939.

With the weather improving, Hitler prepared for a new campaign in the West, the subjugation of France being the ultimate prize. But first he had to guard his northern flank, particularly against any Allied move to secure Norway. This occurred on 28 March 1940 and a counter move by the Germans got underway on 5 April. Using primarily crack parachute troops, the country was rapidly neutralised. Hitler had accepted the Norwegian surrender terms by the 19th. Operation *Weserubung* had the air support of a number of fighter and *Zerstörer* units, including I./ZG 1, I./ZG 76 and II./JG 77.

The first week of May 1940 had a heavy air of expectancy on the Western Front. *Luftflotte* 2 and 4 were poised to spearhead another *Blitzkrieg* without warning. The Germans faced France, Belgium, Holland and Luxembourg, a seemingly impregnable land mass if they contemplated a 'conventional' invasion, for the front was now partially guarded by the massive concrete fortifications of the Maginot Line and Fort Eban-Emael.

Battle for France

Dawn, 10 May 1940. German bombs fell on 47 French airfields including 33 in the most forward Army Zones as medium and dive bombers sortied to gain the vital element of surprise. The bombing was however relatively ineffective: in the French Northern and Eastern Army areas the total loss of aircraft was four and 16 machines respectively. What the attack did achieve was widespread confusion among French forces, a factor that was almost as detrimental to a cohesive defence and counter-attack as widespread material losses.

Kesselring's *Luftflotte* 2 supported von Bock's Army Group B and included the fighters of JG 21 (III./JG 54) and JG 27 plus JG 3, 26 and 51 under *Jagdfliegerführer* 2 to cover the direct thrust into Belgium, while von Runstedt, commanding Army Group A, chose a route through the 'impenetrable' Ardennes forest bordering southern Belgium and Luxembourg. The single-seat fighter support for Army Group A and part of Army Group C was provided by JG 52 and 53.

With the *Panzers* rapidly advancing, the *Luftwaffe* maintained excellent close air support. Bf 109s ranged all across the front, attacking the *Armée de l'Air* wherever its aircraft put in an appearance. Achieving widespread confusion by simultaneously attacking points all along the front rather than confining their sorties to one area, the *Jagdfliegern* ably supported the *Stuka Gruppen* and took on a variety of enemy fighters that rose to challenge them. These adversaries included Hurricanes and Gladiators flown by the Belgians, Dutch Fokker D XXIs and the indigenous French types, the Bloch 155, Dewoitine D 520 and Morane Saulnier MS 406, well supported by the American Hawk 75. Some of these, particularly the Dewoitine D 520, easily the best fighter France had, gave the *Jagdfliegern* a good run for their money although losses for 10 May were to be incredibly low, at only ten Bf 109s.

Over Holland JG 26, 27 and 51 clashed with Dutch fighters which attacked waves of Ju 52 transports landing troops at key points. In five days of bitter fighting, Fokker D XXIs succeeded in destroying a substantial number of Ju 52s and other German types, although there were only 28 of them in service on 10 May. This force, while exacting retribution from the *Luftwaffe*, was inevitably whittled down until only five machines were serviceable when the Dutch capitulated on 14 May.

The attack on the Low Countries hit the French fighter units based in the Northern Air Command particularly hard. Equipped mainly with the Bloch 151/152 series, these groups were to lose more than half their total strength within one week, many to attacks on their bases. Over 100 Blochs were lost in combat by the end of May, although all these were replaced by new deliveries later in the month. French factories had managed to complete a high number of fighters before the German attack, but many of these were non-operational and awaiting delivery from depots after final equipment installation. Disruption of what was already a slow and unwieldy system did nothing to aid the French in their fight.

Luftwaffe fighter operations were primarily aimed at keeping the enemy air forces from hindering the offensive by destroying key points, particularly river crossings; the more natural barriers the *Panzers* passed in the opening days, the harder they would be to repulse without prohibitive loss. Therefore, desperate attempts were made by Allied aircraft to destroy the bridges over the Albert Canal and River Maas. In largely preventing these attacks, the German fighter pilots flew almost continuous dawn to dusk sorties. Coming up against a motley collection of French and British machines, none of which was really suited to ground attack work, or available in adequate numbers for a sustained offensive, the German pilots were often presented with what amounted to sitting ducks. Types such as the Battle, Blenheim, Breguet 693 and Amiot 143 were shot down with consummate ease.

Allied fighters frequently challenged the Bf 109s but there were too few of them to concentrate over the danger points; many pilots had not previously seen any action and their predicament was compounded once the French ground control began to break down. Liaison between the RAF and *Armée de l'Air* also left something to be desired as there had been precious little time to develop a sound defensive strategy using the most capable machines of each air arm to the best effect.

Not that pilot inexperience was solely the prerogative of the Allies – many Germans had little flight time on the Bf 109 and some had not seen action previously. Among those who counted the French campaign as their true air combat debut (despite service in Spain) was Adolf Galland. Serving with *JG* 27, Galland opened his score on 12 May with a surprisingly easy victory over a Hurricane, which he believed at the time to be Belgian. It was later identified, along with two other machines shot down later in the day, as British.

Galland was not alone in expressing the view at de-briefing that these Hurricanes appeared to him to have been indifferently flown, with the pilots not keeping a good look out for Messerschmitts! Although similar criticism could at times be directed at pilots on both sides, this would probably have been hotly refuted by the victims...

The truth was that relatively few Allied pilots had previously had the chance to fire their guns in anger and they persisted in flying formations that could be extremely unwieldy in combat. It was not therefore surprising that the confusion that soon reigned all across the Western Front was compounded by Allied formations being bounced by Bf 109s and taking casualties almost before the pilots were aware of the danger. The multiple French and British aircraft types, each with a differing performance, that had somehow to be melded into a cohesive force was a task that was almost impossible to achieve satisfactorily, particularly under war conditions.

By comparison, the *Luftwaffe* had many advantages not perhaps readily apparent. It was clear that the fighter units had a superb aeroplane and that the pilots had been schooled to execute an excellent set of tactics, flexible enough to handle a rapidly-changing situation – but there was also the advantage of a smaller number of aircraft in other categories. There were three medium bomber types, each with a not dissimilar performance, two dive bombers and a long range fighter. There were other German front line combat aircraft but the *Jagdfliegern* found the Ju 88, He 111 and Do 17 series relatively easy to protect from Allied fighter attack, even if the Ju 87 sometimes gave them a headache in this respect. The Bf 110 crews could look after themselves.

Also, the fighter units were able to rely on a mobile ground support organisation trained to deploy at short notice and to establish basic facilities at forward bases with the minimum of delay.

Integral to this plan was the utilisation of the *Luftwaffe*'s standard transport, the Ju 52, for prompt deliveries of fuel, food, ammunition, spares, mail and the 1001 other items that enabled aircraft to remain operational and made life tolerable for unit personnel. *Luftwaffe* specialists accompanied advance parties of groundcrews and mechanics to quickly establish radio links and communications via telephone and teleprinter so that even a spartan fighter strip could be made operational very quickly. Reliable air transport was one of the less obvious advantages that the *Jagdwaffe* possessed in comparison with the Allies, which were not as efficient in this respect.

On 12 May another German pilot proudly entered the results of his first combat in his log book. Gunther Rall, flying with 8./*JG* 52, had been ordered to escort a reconnaissance aircraft safely back over the German lines. The Bf 109s caught up with it almost at the same time as it was spotted by a flight of Hawk 75s.

Intent on their quarry, the French pilots failed to see the diving Messerschmitts. Rall picked off one and had the satisfaction of seeing it go down for his first confirmed victory. Like many of his contemporaries, Rall had experienced the frustration of hearing about other pilots' success. Getting that sometimes elusive first victory, he later reflected, gave a great boost to a pilot's self-confidence.

When the *Wehrmacht* broke through the French lines at Sedan to split the Allied defence, the successful conclusion of the battle for France came that much nearer. The Germans had had as yet, few major setbacks. In the skies above, the *Luftwaffe* support never faltered – the only slight problem that began to materialise was that the advance threatened to out-range the Bf 109 units. Even though forward airfields could be made operational quickly, the speed of the *Panzers* surprised even the *Luftwaffe* at times!

Increasingly, the *Jagdflieger* came up against the Hurricane, a type able to offer the Bf 109 a dangerous challenge if the RAF pilots knew their business. And more than a few eager young Germans also came to rue the day they came up against French fighters flown by able pilots and those flying the Hawk 75 with its four

.50 in machine guns. The air campaign was hardly one-sided.

But increasingly, the downed German pilot could consider himself marginally safer than his Allied counterpart. With French-held sectors of the front shrinking by the day, there was precious little time to worry about prisoners, although some members of the *Jagdwaffe* were lost 'for the duration'. *Fw* Gerhard Herzog of 2./JG 26 was shot down by MS 406s early in the campaign and captured by British troops, whisked out of France to England and thence to Canada to sit out the rest of the war.

A similar fate might well have befallen Werner Mölders, who was captured near Compiègne early in June. Then the leading Bf 109 *Experten* with more than 20 victories, Mölders had also been the first *Jagdwaffe* recipient of the *Ritterkreuz* (Knight's Cross) on 29 May. The French missed the substantial propaganda coup of removing Mölders entirely from the front and he was, to the delight of his fellow pilots in III./JG 53, subsequently released unharmed.

German fighter losses, while appreciably lower than those of their opponents, inevitably included machines flown by pilots who had already made a name for themselves. On 21 May Dr Erich Mix, then a *Hauptmann* in III./JG 2 was shot down at Roye and posted as missing. *Oblt* Paul Gutbrod of II./JG 52 who had flown the Bf 109D in some of the earliest combats over France, was killed on 3 June. Typical of the confusion over missing aircrew that affected both sides, Mix was soon found, very much alive.

By late May, France's military position was dire. The remnants of the French Army and the British Expeditionary Force were retreating to the coast, where evacuation from the continent to England appeared to be the only way to continue the fight. In time, despite all the *Luftwaffe* could do to stop it, the evacuation from Dunkirk, a defeat for the Germans of not inconsiderable proportions, saw 330,000 men whisked away.

It was over Dunkirk that the *Jadgwaffe*'s Bf 109s met RAF Supermarine Spitfires in combat for the first time. Fresh and spoiling for a fight, the British pilots were occasionally surprised at the ease with which they could shoot down the much-vaunted '109'. For the Germans, the Spitfire became something of an icon; pilots who met it in combat realised that here, for the first time, was a fighter good enough to take on their *Emils* on equal terms.

It was probably over Dunkirk that the *Jagfliegern* began the unconscious myth of 'Spitfire snobbery'. From then on, almost every German pilot fired on by a fighter with RAF markings was 'attacked by Spitfires'. In truth the marked difference between the Hawker and Supermarine fighters could be hard to distinguish in the melee of air combat, particularly during these early engagements. Many a Bf 109 pilot, intent on hauling his machine out of the hosepipe fire of eight Browning machine guns, watching out for his wingman and the enemy to avoid a collision, scanning the sky for the bombers he was supposed to be protecting and keeping an eye on fuel state, engine revolutions, temperature and so on, all at the same time, could be forgiven for missing the finer recognition points of Hawker or Supermarine design! But the fact remained that after Dunkirk, few German pilots seem ever to have been shot down by Hurricanes!

Bf 109 pilots also made a recognition error in the case of the Boulton Paul Defiant. Making its combat debut over Dunkirk, this turret fighter looked very similar to a Hurricane from some angles and a number of *Jagdfliegern* were to pay dearly for their mistake when fire from the four Brownings caught them unawares. The Germans did not however, make too much of a habit of coming into range of the upper turret and instead concentrated in attacking from below or from directly ahead, where the Defiant's guns could not track them. Crews of the ill-fated Defiant squadrons were soon to realise that their 'secret weapon' was not half as effective as they had been led to believe.

Fighter combat over the Dunkirk beaches fell heavily on the shoulders of pilots of most of the front line *Jagdgeschwadern*, including JG 2, 3, 26, 27, 52 and 53. Pilots belonging to all these units found that trying to protect bombers that were in turn, attempting to prevent Operation DYNAMO from succeeding was a wearying and near-impossible task after weeks of combat over France.

Bombing troops dug into soft sand which cushioned the blast effect achieved little and the fighter pilots began finding out just how difficult it was to cover bombers subjected to determined attack by fighters that were more than capable of destroying them. Most demanding of all in this respect were the Ju 87s. Slow in level flight, the *Stukas* were difficult to keep station with but when they pushed over into their steep-angle dive, they would leave the escorting Messerschmitts standing. All the Bf 109 pilots could then do was to watch for their comrades' machines to pull out and level off, when they were at their most vulnerable. By maintaining a high cover, the Bf 109s could often dive to the rescue, but *Stuka* escort remained one of the most exacting tactical tasks the *Jagdfliegern* ever had.

At the time, the confused air operation over Dunkirk only compounded the existing problem of fighter pilot claims versus actual losses. These latter were usually lower than those claimed and although the unadjusted figures might have created some satisfaction in London and Berlin, the RAF had to conserve fighters for whatever direct action against England the Germans were planning. For the *Luftwaffe*, the French campaign had been fairly costly in terms of bomber losses, even though the number of Bf 109s removed from strength was not significantly high and replacements in terms of both pilots and aircraft soon made good any shortfall.

Tangling with the RAF over Dunkirk however, was when the *Jagdwaffe* experienced its first losses over the English Channel, where the rescue and return of wounded personnel was that much more difficult than it had been over France. When combat over the Channel resulted in reports of pilots missing, those unfortunates would, in some cases, never return. Not that this period of fighting was prolonged, or that the German fighter pilots had lost any of their prowess. As an illustration of relative losses at the latter stages of the French campaign the casualties for 1 June 1940 totalled eight Bf 109s from JG 26 and one from I./JG 20 (the outmoded unit designations of *Luftwaffe* units continued to appear in casualty returns throughout the battle of France). This compared favourably with claims for ten Spitfires (from Nos 19, 41, 222, 609 and 610 Sqns) and eight Hurricanes (Nos 17, 43, 73, 145 and 245 Sqns).

Away from the action at Dunkirk, the *Wehrmacht* was completing the humiliation of the French. Reims was secured on 9 June and less than a week later German troops were marching down the *Champs d' Elysee*. Having been declared an open city, Paris was spared the devastation of other European capitals, and was taken without heavy fighting.

With Paris in German hands, the fate of France was all but sealed. With relative ease the *Panzers* broke directly through the fortifications of the Maginot Line and secured historic Verdun and the port of Cherbourg within three days of each other. Finally on 22 June, the French Government sued for peace.

For the *Jagdfliegern*, the incredibly successful air campaign had virtually finished on the 19th, there being no further Bf 109 losses as a direct result of enemy action over France after this date. To help secure France, the *Luftwaffe* lost 1,469 aircraft up to 25 June 1940; 1,009 German and Italian machines were officially credited as destroyed by the *Armée de l'Air, Aéronavale* and anti-aircraft units, the total including 355 Bf 109s and 110s. Of these, approximately 180 Bf 109s were destroyed or written-off as a result of air combat, collision and accidents where the aircraft sustained more than 50 per cent damage.

Among the statistics of fighter versus fighter combat where the opposing aircraft type was identified, *Luftwaffe* records indicate that seven Bf 109s were shot down by the Fokker D XXI over Holland during the first two days of fighting. The Bf 109 had, not surprisingly, been responsible for the highest percentage of French fighter losses, the final figures including approximately 100 D 520s from an overall loss figure of 250 fighters of all types lost in action. RAF fighter losses, (Hurricanes and Gladiators), totalled 219.

For what the data was worth, the French and British had had ample opportunity to examine the Bf 109E even before the war in the West had begun. Evaluation of captured examples had been possible from 1937 and both countries drew up comprehensive performance figures to give a fairly reliable impression of the plus and minus factors of the D 520, Hurricane and Spitfire versus the Bf 109E. That such data can only be of limited value if it cannot be exploited to the full by sound tactics, organisation and adequate numbers, was surely shown by the outcome of the battle for France.

Even though large areas of France had not been overrun by the Germans, the speed of the advance had taken the *Wehrmacht* from the border to the Channel coast in less than six weeks. German flags flew in Paris; the heady pace of it all could scarcely be believed in Berlin – and yet here, undeniably, was the prize. In one of the most rapid and successful military campaigns in history, Hitler had finally wiped out what he perceived as the shame of Versailles.

For German airmen, the armistice brought less heady celebrations. It was a chance to catch up on sleep without a dawn-to-dusk alert, continual operations for days on end with the ever-present risk of death or injury and the perpetual moves to new airfields and living quarters. For some weeks the efficient German military machine wound down. Units of the *Jagdwaffe* were gradually moved forward, some to occupy mere strips of flat grass as close to the Channel as possible, others to enjoy the more permanent facilities of established former French airfields. New fighters and personnel arrived to make good the attrition of the past weeks but overall the force was in excellent shape and eager for the next round.

Hermann Göring offered Hitler expansive assurances that his air force could create the necessary conditions for an invasion of England to take place; paramount among these was the destruction of the British fighter defence and publicly at least, the *Luftwaffe* C-in-C appeared to make little distinction between the English Channel and the Albert Canal. Yet the fact that he was proposing a totally new military venture without any historic parallel appeared to go almost unnoticed. In short, he was preparing to commit a tactical force to a strategic role – which was if nothing else, novel. Another factor that would be lacking in a war on England was the element of surprise: the defences would be fully alerted as to the *Luftwaffe*'s intentions and in this particular case, able to offer a strong defence.

But the euphoria of the French defeat swept away all possible doubts as to the *Luftwaffe*'s ability to perform such a gargantuan task and many individuals in the German High Command lacked the knowledge to realise that the *Luftwaffe* chief had airily glossed over the glaring and highly significant differences between an attack on England and the land-locked countries of Europe. Göring all but convinced Hitler – and possibly even himself – that the Channel was no more of an obstacle than a wide river. And the *Luftwaffe* had flown across many rivers . . .

Chapter 4
Channel Clash

On the face of it, the English Channel did not seem to the average *Jagdflieger* to be an insurmountable obstacle. At its narrowest point it separates Cap Griz Nez from Dover by a mere 20 miles. It was only when the single-seat fighter pilots got down to making some simple calculations that they came to appreciate the difficulties that that short stretch of water could present.

Given that the British would fight, German pilots would have at best ten minutes' combat time over southern England before having to break off and return to base to refuel. Even though in modern air combat terms ten minutes was a lifetime, it unrealistically presumed a Bf 109 in perfect condition, one not having to waste fuel searching for the enemy, flown by a pilot fully alert and able to spot opposing aircraft early enough and preferably one totally unhampered by a headwind or other adverse weather condition. Allowing time for any one of these factors, German fighter pilots could be down to dangerously slim fuel reserves very quickly indeed.

The Bf 109E's fuel system included a cockpit instrument panel warning light which illumin-

Among the victims of the *Kanalkampf* was *Oblt* Hans von Werra, Technical Officer of *JG* 3 who crash-landed his E-4 in Kent on 5 September 1940. (*Kent Messenger*)

Tactical camouflage, an overspray of the Bf 109E's *Hellblau* fuselage and undersides, was introduced on German fighters during the latter phases of the *Kanalkampf*, as 'yellow 5' shows. *(via R L Ward)*

ated when fuel dropped below 200 litres – or was supposed to. The contents of the tanks could vary between individual aircraft and the actual fuel level was sometimes considerably lower than the gauges suggested, a fact that hardly increased a pilot's peace of mind!

Whether or not the RAF would venture over France was a question which could not be answered – although the strength of the *Luftwaffe* appeared to rule out this possibility. That meant that the *Jagdfliegern* could expect to meet the enemy only over the waters of the Channel, or England itself. Neither was a very encouraging prospect if pilots were shot down or had to bale out. On the positive side, incarceration in England would most surely be brief, as captured airmen could look forward to early liberation by invading German forces – just as they had in Poland, France and the Low Countries . . .

The limited range of the Bf 109E was to be an increasingly important factor in this new phase of operations; strangely, the *Jagdwaffe* did not receive supplies of external fuel tanks for its *Emils*. These had been used on front line aircraft in Spain and notwithstanding individual pilots' dislike of such tanks, due to the inherent fire risk (one of the earliest types developed for fighters was constructed of wood) and the additional weight, with consequent restriction on manoeuvrability, it was late 1940 before the Bf 109E began to carry a 300-litre drop tank more or less as standard to boost its meagre 400 litres (88-gallon) internal tankage.

Luftwaffe pilots' supreme confidence in themselves and their trusty *Emils* was one thing but there was little that could immediately be done to extend an endurance of about 55 minutes with the aircraft flying at medium altitude (16,400 ft) and using maximum continuous engine revolutions (2,400 rpm). These figures enabled the Bf 109E-3 to attain a true airspeed of 323 mph and a range of 286 miles. Both could be substantially improved upon (up to 1 hr 50 minutes and 413 miles) if conditions allowed 'maximum economy' engine power settings to be used for any length of time.

The favoured and highly effective *frei Jagd* patrols – where the *Jagdfliegern* had freedom of action – would be the order of the day over the Channel and the young 'Turks' enthused at length at what they were going to achieve in the coming battle. Some rashly swept aside the undoubted qualities of the Spitfire while others could hardly wait to test their skills against the best fighter the Bf 109E had yet faced in this war. Some were content to await events.

Throughout June preparations for the air assault on England went on until a sizeable fighter force had occupied forward airfields, mainly in the Pas de Calais area. On 30 June and 1 July respectively the Germans invaded the Channel Islands of Guernsey and Jersey and for a time, a small number of Bf 109Es of JG 53 could claim the unique distinction of being the only German fighters to be fuelled from British bowsers driven by British personnel. Otherwise based at Dinan and Cherbourg, the component *Gruppen* of JG 53 used Villiaze on Guernsey and to a lesser extent St Helier on Jersey, as forward bases during *Kanalkampf* operations. The island airfields proved to be a safe haven for a number of Bf 109s, particularly those which had sustained combat damage. Part of JG 53 remained on Guernsey until the spring of 1941.

Each component *Gruppe* of the main *Jagdgeschwadern* had been brought up to strength and generally re-equipped with the Bf 109E-3 and E-4 models, although many examples of the older E-1s remained in service. Even though the latter lacked wing cannon, its significantly lighter all-up weight compared to later models was seen as advantageous by some pilots, who viewed any weight increase, no matter how sound the reason for it, as anathema.

Others noted with some satisfaction that the cannon firepower of the E-3 and E-4 gave the average pilot an even chance to extricate himself from a tight spot if he happened not to be the best of marksmen. Some pilots who strove hard to emulate the *Experten*, found that a respectable tally of aerial kills could be frustratingly hard to push beyond single figures, so a good spread of fire was more than acceptable.

Neat line-up of Bf 109E-4s, probably in France *circa* autumn 1940, with 'white 9' nearest the camera. (*via Robertson*)

Testing the *Friedrich*

Even before the *Kanalkampf* had entered its decisive phase, Messerschmitt A.G. had completed a significant reconfiguration of the Bf 109E. This far-reaching programme resulted in a new, more aerodynamically efficient and streamlined fuselage which while retaining its pedigree, swept away most of the drag-inducing features of the *Emil*. Installation of the DB 601E engine of 1,300 hp to power the Bf 109F had also offered the opportunity, not only to fair smoothly the contours of the cowling but to entirely redesign the access to the engine. The skin of the forward fuselage nose was made in two panels which were hinged along their top edge to expose both sides of the engine above the exhaust ports, while the entire lower nose section carrying the main coolant radiator fairing, was made to hinge down and to the side. Whereas most of the Bf 109E's cowling panels had to be removed entirely, those of the F introduced a series of Zuiss fasteners which held each of the three main sections in place when they were closed. Retaining stays were provided so that most 'in situ' servicing could be carried out without removal of panels. This feature, by no means common on fighter aircraft of the period, undoubtedly helped endear the 109 to *Luftwaffe* ground crews. In the field, the minutes saved undoubtedly made the difference between a timely '*Alarmstart*' (scramble) and a delay on many occasions.

And Messerschmitt's fighter now exemplified a philosophy, widely accepted in some quarters, that air combat results were achieved far more positively by the pilots aiming the nose of the machine at the target without having to allow for the harmonisation of wing guns. Not everyone agreed with this view and although the Bf 109F was considered a vast improvement on the E aerodynamically, it was criticised as lacking in firepower, particularly with the pilot of average ability in mind.

The first Bf 109F conversion made its maiden flight on 10 July 1940, this event being recorded by an airframe that began life on the production lines as a Bf 109E and fitted with the DB 601E. The revised engine cowling, cantilever tailplane and a retractable tailwheel further distinguished the new model, these improvements having been progressively tested on the Bf 109V17 and V18, both of which were also fitted with a new wing. Incorporating extended wingtips with a rounded profile, the Bf 109F wing was structurally different from that of the E model, particularly in having re-profiled radiators and dispensing with the bays for integral armament.

Production of the Bf 109F-0 and F-1 was initiated during the summer of 1940 in time for some examples to be issued to front line units by the late autumn. Operational flying revealed that the elimination of the Bf 109F's tailplane bracing could, under some flight conditions, impose critically high loads on the entire structure and lead to the rear fuselage breaking away.

These technical challenges lay a few months in the future as the Bf 109E-equipped *Jagdgeschwadern* made ready to annihilate RAF Fighter Command. When *Luftflotte* 2 was established in north-eastern France, Holland and Belgium and *Luftflotte* 3 in north-western France the combined strength of their subordinate units exceeded 1,000 aircraft, 760 of them single- and twin-engined fighters.

Despite a relatively high number of operational *Gruppen*, the *Luftwaffe* of 1940 remained something of an exclusive club; many of the commanders of fighter, *Stuka* and bomber units not only knew their opposite numbers in other

Jabo operations required some ground support equipment including bomb-handling trailers. A 250kg SC bomb is being loaded in this view. (*Robertson*)

Romania was among the Axis nations that flew the Bf 109E operationally, two of her E-4s being shown here in formation with a German machine.

Geschwadern but often counted them as close friends. This was particularly true in fighter units, where good liaison and regular conferences disseminated all the latest news on tactics and equipment. Wide-ranging discussion of the latest intelligence on the enemy's potential ensured that everyone was fully informed.

By mid-1940, many German fighter pilots had already pitted their Bf 109s against a diversity of opponents in aerial combat and few surprises were expected from the RAF; both Spitfires and Hurricanes had been shot down, the former in far fewer numbers admittedly, but neither had thus far been fitted with cannon, according to the latest reports. Unless the British had something else that they had kept a dark secret, the Messerschmitt pilots would surely prevail. Much would depend on tactics, pilot skill and numbers.

There were persistent rumours that Fighter Command had an early warning system to detect aircraft approaching the coast and the *Jagdflieger* realised that this could prove troublesome. The element of surprise that had immeasurably aided German air operations before was of course, gone. Only time would tell how the British radar system would stand up to pounding by the *Kampfgruppen* – for this should be a priority target.

Preparations for the great air assault only gradually gathered momentum; when convoy attacks began on 10 July, the Bf 109s found themselves flying cover to *Stuka* formations and medium bombers attacking Channel shipping and English coastal ports, with mixed results. A compensation for them was the *frei Jagd* sweeps which brought contact with British fighters. The enemy appeared reluctant to combat Messerschmitts and if bombers were not present to draw the fighters, the *Jadgfliegern* occasionally found little action.

The speed and direction of RAF fighter interception impressed the Germans however and it became clear after a period of assumption that standing patrols were being mounted, that it was radar early warning that enabled accurate intercept vectors to be passed to the fighter squadrons. But a disadvantage of the 'Chain Home' radio-location finding (RDF) system, unbeknown to the Germans, was the fact that while it gave basic height and direction information, it could not differentiate between fighters and bombers.

That Fighter Command was being well served by radar continued to be only a suspicion on the part of the Germans and what they never were to know was that Churchill and Air Marshal Hugh Dowding were actually reading transcripts of their own signals via the *Ultra* network. Even in its infancy this early penetration of the Germans' secret *Enigma* radio traffic encoding machine enabled much useful, if at times fragmentary, data to flow into Fighter Command HQ at Bentley Priory.

For their part the *Jadgflieger* had only to concentrate on the task in hand – the destruction of Spitfires and Hurricanes in such numbers that they could no longer harry German bombers. Without any ground dimension, the Messerschmitt pilots found the '*Kanalkampf*' very different to any theatre of war they had yet experienced and previously-held operational beliefs began to be revised in the face of

A Bf 109E-4/B operated by the *Jabo Staffel* of *JG* 26 taxying out with the aid of a groundcrewman and carrying four 50 kg bombs under the fuselage.

determined fighter defence. The vulnerability of the Ju 87 did not, perhaps, come so much of a surprise as did the relatively poor showing of the Bf 110. Often out-manoeuvred by the RAF single-seaters, the *Zerstörer* crews were now unable to exploit their heavy battery of cannon.

Flying their 'coat trailing' sorties off the coast, the Bf 109 *Jagdgeschwadern* observed that the British appeared not to have developed a heavy AA defence, and that many potential targets were covered by barrage balloons. This 'passive defence' could be hazardous if aircraft ran into the heavy balloon cables and when there was otherwise little action, the *Jagdflieger* delighted in shooting the elephantine monsters down, which if nothing else, provided good target practice!

But the *Jagdwaffe* soon found that it was taking a spiralling number of fighter sorties to destroy a handful of the enemy and even as early as mid-July, the daily pace of operations was surprisingly intense. As well as combat flights across the Channel, there were local air tests to be flown, new pilots to familiarise in operational matters and some ferrying of Bf 109s between the French airfields upon which component *Gruppen* of each *Jagdgeschwader* were based. The July weather was generally cloudy, with considerable rainfall, which necessitated the rigging-up of extra shelter for servicing on the many airfields that lacked such facilities under cover. All *Gruppen* had to have fighters at readiness throughout the daylight hours.

By July the preparations for Operation *Seelöwe* (Sea Lion), the invasion of southern England in line with Hitler's *Führerweisung* (Directive) No 16 issued on the 16th, had not been advanced significantly insofar as air superiority was concerned. Bombs had been dropped on coastal areas, but in the main the *Luftwaffe* had thus far concentrated on shipping, despite the prominent latticework towers of the RDF stations offering perhaps more worthwhile targets. Although German photo reconnaissance gave some indication of the damage done, there remained the difficulty of accurately assessing the results.

The tactical bombers of the *Kampfgeschwadern* had not by then lost a large percentage of their number but it was clear that individually the He 111, Do 17/215 series and Ju 88 could be destroyed by the RAF fighters with ease. None of the bombers were themselves heavily armed; all defence consisted of small calibre, hand-held machine guns, there being a total lack of powered gun turrets. Bomb loads were also modest, leading to a paramount need for formations to remain integral if any lasting damage was to be done to their selected targets.

Having never been subjected to sustained attack by fast, well armed fighters, the hapless *Kampfgeschwader* crews found the war against England a traumatic experience. They had little choice but to adhere to their tactics, such as they were, and to accept some of the accolades, most of which had previously gone to dive-bomber crews, for their new role. The problem was that the bomber crews had been trained for a tactical war in support of the army and they had hardly practiced close formation-keeping during independent operations, nor had they much experience of providing mutual gunnery support under fighter attack. Thus were the broad parameters set as the *Luftwaffe* geared up for another lightning military success . . .

Göring's timetable for the complete subjugation of the RAF was a highly optimistic six weeks. He reasoned that as a country the size of France had succumbed to the German war machine in that time, England should hardly take any longer. On paper, the number of aircraft available to the opposing sides did not give the RAF much chance against the hitherto invincible might of the *Luftwaffe*.

The *Jagdflieger* initially did their best to wear down the RAF to the point that when the High Command judged the conditions to be right, the bomber crews could expect a fairly easy ride. The problem was that very few Germans had any idea of the true strength of the enemy. British fighters actually shot down into the Channel (i.e. seen to go in by one or more witnesses) were duly entered in pilot combat reports and the numbers conscientiously tallied on a daily basis and forwarded to *Luftwaffe* HQ in Berlin. But nobody knew how crippling the British losses were, how quickly front line losses could be replaced, what sort of organisation existed for the recovery and repair of damaged machines, or what the output of the factories was.

Above all, it was all but impossible for the German flyers to gauge whether an enemy fighter pilot taking to his parachute would be fit enough to fly another aircraft against them within hours, days – or at all. Pilot injury was perhaps the 'x-factor' of the battle for England and one that held quite a significance for both sides.

What the *Jagdfliegern* did experience was a pattern of their own individual losses that became irksomely repetitive. Lucky individuals were those who, obliged to ditch their Bf 109s in the Channel, did so under some semblance of control, managed to extricate themselves and swim away, buoyed up by a kapok-filled life-jacket or preferably, aboard the dinghy that every pilot carried as part of his survival kit. A timely radio call by their *Staffel* colleagues would alert the ASR aircraft of the *Seenotdienst* or a rescue boat stationed on the coast by the navy and, barring further mishap, the downed pilot would be returned to his unit without further delay.

The other side of the coin was that pilots simply began to go missing. If they had come down in England and been made prisoner, the German authorities were notified, albeit some time later. Some badly wounded men were repatriated during the war but most did not return to Germany for the duration of the war.

The *Jagdflieger* were however, inflicting more damage on the RAF than they were losing in terms of Bf 109s. By the end of July the fighter units on the Channel had destroyed 35 Hurricanes and 37 Spitfires, plus six Defiants and five Blenheims for the loss of 48 Bf 109Es. German bomber losses had been somewhat heavier.

Terror bombing, i.e. scattering a small weight of bombs over a wide area to cause panic among the civilian population, had been successful elsewhere but there were few signs that it was achieving anything of the kind in southern England, for the inhabitants knew full well that no German tanks would appear on their beaches without due warning. Hitler's avowed intention of subjecting Britain to a full scale assault designed to ruin the country's economy and starve its people, was far from being achieved.

That is not to say that the *Kampfgeschwadern* had not carried out any damaging attacks; RAF airfields had begun to be heavily bombed and considerable damage had been done to aircraft and installations. But here again, German intelligence was initially unable to prioritise the type of airfield attacked. Training and bomber bases and Fleet Air Arm shore stations which did not house fighter squadrons were attacked along with the main Fighter Command airfields, although the latter soon came to be identified for what they were.

Luftwaffe commanders had planned a grandiose opening attack – albeit some weeks after the battle had from the British viewpoint, already started – as the prelude to a series of knock-out blows to the RAF. A postponement due to inclement weather, put this back to the second week of August. Before that the *Jagdwaffe's* task was growing progressively more demanding; the RAF fighter squadrons, although they were being hurt and the shortage of pilots was a cause of great concern, invariably rose to intercept German bombers and if it was totally unavoidable, to dogfight the Bf 109s.

Adler Tag

So confident was Hermann Göring that his *Luftwaffe* would have little difficulty in simply carrying out his orders that he travelled to the French coast to watch the bomber force begin the mass assault that would ultimately finish England's will to resist. It was 13 August and the thunderous sound made by waves of He 111s, Dornier 215s and Junkers 88s was probably music to the C-in-C's ears. Soon the ever-alert Fighter Command raid plotters detected a huge mass of bombers en route to England. *Adler Tag* (Eagle Day) was unfolding . . .

Having been informed that almost the entire strength of *KG* 2, 26, 27, 30 and 54, plus the Ju 88s of *LG* 2 and Ju 87s drawn from *StG* 2 and 77 comprised the bomber force that morning, the German fighters had, in different sectors, a sizeable force of bombers to protect. *Luftflotte* 2's fighter element that day was composed of the *Stab* and all three *Gruppen* of JG 3, 26, 51 and 52, plus the *Stab* and I./JG 54. *Luftflotte* 3 provided the *Stab* and three *Gruppen* each of JG 2, 27 and JG 53.

Adler Tag was intended to be the opening

round, in line with Hitler's *Führerweisung* (Directive) No 17, for intensified air and sea warfare against England. Accordingly, *Luftwaffe* sorties on 13 August totalled a record 1,485 – and 45 aircraft were lost in return for 13 RAF fighters, but with only three pilots killed and two wounded. Five Bf 109s and eight Bf 110s were also lost. Clearly, this was not what was required . . .

On the 15th, a missed recall order was to result in some bombers arriving over their targets with no fighter protection: all three *Luftflotten* were briefed to attack RAF airfields and bombs fell on Driffield, Worthy Down and Middle Wallop, as well as the more important stations at Manston, Hawkinge, Martlesham Heath and Rochester, home of Short Bros. RAF fighters put up 974 sorties and destroyed 75 German aircraft. The *Kampfgeschwadern* had again been briefed for too many targets and their bombs were not well concentrated on any one objective. *Frei Jagd* sorties were flown later in the day, but the presence of the Bf 109s could not prevent the majority of the bomber losses.

This was not the kind of loss ratio that was going to give the Germans air superiority, as their own bomber and fighter casualties were invariably increased, if only marginally, by non-operational accidents which damaged or wrecked aircraft and injured crewmen. Fighter pilots suffering such mishaps would be off the duty rosters for days, if not weeks, the inevitable toll from operational attrition that reduced the number of experienced flyers available for the daily sorties.

The *Jagdflieger's* task grew ever more demanding as calls for bomber escort grew proportionally with the scale of RAF reaction to raids. Most of this duty fell on the shoulders of the *Jagdgeschwadern* for if there was one factor to encourage the Germans in this phase of the air war, it was the performance of the Bf 109 units. While they continued to fly the *frei Jagd* sorties which were so damaging to Fighter Command, they were increasingly obliged to provide close escort to bomber formations, which robbed them of their vital flexibility. Waiting for the enemy to attack went against the grain and nullified their excellent tactics which were based on retaining freedom of action at all times. Ludicrously, the 109 pilots also found themselves shepherding *Zerstörern* when the much-vaunted Bf 110 was found to be far more vulnerable to fighter attack than had been realised.

When *Adler Tag* failed to result in the destruction of the RAF, Göring was incensed. Blind to the true facts of the operation, he appeared to see only that a personal guarantee to his *Führer* was not being fulfilled. The answer, as he saw it, was to shake up the fighter force, to find scapegoats for a badly flawed operational plan and to make personnel changes.

On the evening of the 15th, Göring held a conference at which he stated that fighter protection for dive bombers was to be stepped up to the ratio of three fighter *Gruppen* to one *Stuka Gruppe*. He ordered that one *Jagdgruppe* should stay with the Ju 87s and dive with them, one had to maintain a medium altitude watch for enemy fighters while the third established a top cover to ward off fighter attack. All *Gruppen* were tasked to provide an escort for the *Stukas* on their return home.

At a further conference at Karinhall on 19 August the *Oberbefehlshaber der Luftwaffe* (Luftwaffe Commander-in-Chief) ordered that the 'complacent' older *Gruppenkommandeuren* and *Geschwaderkommodoren* be replaced by younger blood, men who could and would achieve his goal and restore his standing with Hitler. Further delay in the inevitable triumph of German arms was unthinkable . . .

As a positive result of this unwarranted criticism of their performance, the *Jagdwaffe* saw a number of brilliant pilots and tacticians elevated to head the cream of the fighter *Gruppen* on the Channel coast. They included Gunther Lutzow (JG 3); Hannes Trautloft (JG 54) and Wolfgang Schellman (JG 2). Among the new *Gruppenkommandeuren* were Walter Oseau, Edu Neumann, Otto Bertram and Gerd Schopfel.

By late August, a further part of the *Luftwaffe* C-in-C's revitalisation plan had worked through to the front line airfields. Now the Bf 109s were covering the bomber forces from every conceivable angle – but the latter continued to be attacked and shot down. The weary *Jadgfliegern* did their chief's bidding and on certain days they did indeed prevent bomber casualties. They were also continuing to whittle away at Fighter Command, which had to contend with heavy raids on the forward sector stations, some of which were virtually removed from the order of battle until repairs could be effected. Resident squadrons were moved out in some cases to operate from unsuitable flying club fields; more potentially damaging to a well-conceived, integrated defensive system was the death of skilled controllers on the airfields but the plotting room personnel were also evacuated and established in makeshift premises where they were relatively safe from German bombs.

Still the Messerschmitts took off two, three, four times a day, formed up and headed out across the grey, forbidding waters. Just as the English blessed this natural moat around their castle, in their turn the German pilots came to hate the very sight of the Channel. Watching their bombers, the *Jagdfliegern* waited for the pencil thin line of Dover's cliffs to bisect their windscreens and their neck muscles began to ache as they scanned the sky for the tiny dots that would materialise as the brown and green camouflaged fighters with 'peacock's eyes' roundels. They knew full well that some of the men flying those fighters were as good as their *Experten* and that their aircraft flew as well as if not better than the *Emil*, which was such a short time ago, master of Europe's skies.

As England came clearly into sight, someone would transmit the alarm and all hell would break loose. If their luck was in, the challenge would come from old Hurricanes flown by novices, perhaps foreign pilots who demonstrably had had little practice in formation flying, R/T procedure or battle tactics. On the other hand it was observed that some of the English flyers were copying their own 'finger four' these days. It really was too much . . .

If the lucky charms were not working, the danger would come from Spitfires flown by skilled and aggressive old hands, the 'English Lords' as they irreverently nicknamed the RAF pilots who, some believed, were drawn exclusively from the ranks of the aristocracy. . . . Then the old *Emil* might have to work overtime even to survive the onslaught, as the Spitfire could turn as well and was more manoeuvrable than the 109 low down. A lightweight designed to fight at altitude, the *Emil* was not at its best below 10,000 feet, where the wing became less aerodynamically efficient. Wrestling with stiffening-up controls, pilots frequently criticised the tendency for the leading-edge slats to deploy, making the machine even harder to control. It was a minus factor that they learned to live with, for their aircraft had much to commend it, including heavy firepower and towards the end of the *Kanalkampf*, increased armour protection.

After combat with Hurricanes or Spitfires, or both, the Bf 109s reformed and flew back to France, cursing the inevitable straggling bomber which some of them would have to shepherd at least as far as the coast. Not being able to communicate that fact to the bomber crew meant that the Bf 109s had to approach cautiously, otherwise the jumpy gunners might let fly.

Making home base without further mishap, there came the sobering count of the casualties, the familiar faces that would not be returning unless they had diverted elsewhere or been picked up, and the almost inevitable sighting of the machine flown by so-and-so diving vertically into the Channel. Another empty place at the dinner table tonight . . .

The pace affected *Experten* and novice alike; individuals like Helmut Wick, Hannes Trautloft, Adolf Galland and Werner Mölders – who was promoted to *Oberstleutnant* (Lieutenant-Colonel) and given command of JG 51 – fully appreciated the problems faced by the inexperienced men, thrown into what was undoubtedly the *Jagdwaffe*'s (fighter force's) toughest campaign yet. Cool and decisive in action, Mölders' stature, already high among his contemporaries and superiors, grew apace. His pilots strove to emulate the achievements of 'Vati' for he was always available to share the benefit of his experience and smooth the path to another's first victory, which he knew was a very important milestone. But he always advised that this had if at all possible, to be achieved without too much of a shock. Striving to prove they could be a useful member of a fighter *Geschwader*, young pilots were prey to over-confidence and relaxation of vigilance that could prove fatal.

Having enemy aircraft kills confirmed was none too easy; the *Luftwaffe* imposed a rigid set of rules under which there would be no credit unless a colleague saw the enemy aircraft fall. More than a few *Jagdfliegern* shot down enemy aircraft only to have the claim officially denied when the circumstances of the combat were scrutinised. Unlike the RAF, the *Luftwaffe* did not allow percentage shares in victories when two or more pilots claimed the same victim – a man either scored a definite kill or not. In cases where there was some doubt as to who had fired the fatal burst, the victory was credited to the *Geschwader* rather than an individual.

Nor was the *Kanal* front a place for the 'lone wolf'. Teamwork was the key to survival and by maintaining a tight system of leader/wingman, the *Jagdflieger* who might otherwise have been shot down, often emerged victorious from a difficult situation. There were exceptions to this rule when pilots had sometimes to fight their way out alone. In that case, claims for personal victories could not be taken at face value and a number of men bemoaned the fact that they were not credited with enemy aircraft they knew they had destroyed.

Any despondency on the part of the Germans might have surprised RAF fighter pilots, for as August waned, their situation grew increasingly critical. Combat operations on the 31st saw the Bf 109s destroy 39 RAF fighters while the Germans themselves lost 33. By that date *Luftwaffe* fighter casualties had risen to 254. Yet while the enemy was losing pilots and aircraft at a rate that could not continue indefinitely, the *Jagdwaffe* was now despatching up to 60 fighters to cover 15-20 bombers.

Apart from attacks on radar sites and airfields – some of which were highly effective – the *Kampfgruppen* were still being briefed for too many targets in south-east England, which invariably spread their efforts too thinly. On 3 September however, orders indicated a significant switch. That day the target for the greater part of the available force was to be London. The bombers also attacked the capital in daylight on the 5th and the 7th, such raids imposing an enormous demand on the *Luftwaffe* fighter force, which had on the latter date, to provide no less than 648 Bf 109Es to cover 625 bombers.

While provision of a sizeable escort was well within the *Jagdverbande's* resources, shielding the bomber stream heading for the capital reduced *frei Jagd* sorties – and resulted in a small but very welcome respite for the forward RAF fighter stations. Given such a breathing space from bombing, the RAF could bring fresh fighter squadrons down from the north and Midlands, carry out some rapid repairs and generally patch together a decently-sized force with which to meet the next onslaught. Consequently, this phase reached something of a crescendo on 15 September when the bomber attack on London was all but routed by upwards of 250 Spitfires and Hurricanes, which shot down 56 bombers.

The daylight attacks on the city continued but by the end of September the bombers were increasingly being briefed to take off from their French bases late in the day and to arrive over London at dusk. This not only minimised their vulnerability to British fighter attack, but largely freed the *Jagdflieger* from escort work. Neither side yet had many fighters suited to night interception.

Monday, 30 September was a black day for the German fighter force. Elements of seven *Jagdgeschwadern* escorted bombers from nine *Kampfgeschwadern* and also flew *frei Jagd* sorties to draw enemy fighters away. The main target was again London although bombs were to fall on other areas of south-east England. Nineteen RAF fighter squadrons were to see action before the day was out.

At 0900 some 200 Dornier 17s, Ju 88s and He 111s crossed the Kent coast en route to the capital. Escort was provided by JG 2. This force was intercepted by eight squadrons of Hurricanes and four of Spitfires. Combat was joined over Maidstone, the British fighters succeeding in turning the bomber formations away from their intended course. Individual engagements resulted in RAF claims for four Bf 109s and four Ju 88s against three Hurricanes.

The first three Bf 109s to fall were all E-4s from II./JG 2 and all the Hurricanes were, coincidentally, from No. 229 Sqn. As a direct result of combat II./JG 2 lost a further two aircraft in a take-off collision at Octeville airfield, both pilots being killed. Another II *Gruppe* pilot was able to walk away from a minor landing accident at Calais-Marck on his return from combat. A machine belonging to 9./JG 26 was the fourth Messerschmitt lost in action. The pilot was killed when his machine dived into the waters off the Thames Estuary following combat with No. 66 Sqn's Spitfires.

A series of *frei Jagd* sorties were mounted by JG 2 and 9./JG 53, plus the Bf 110s of ZG 26, although these forays achieved little. A second attempt to break through the British fighter screen to bomb London was made at midday with a force of 100 aircraft covered by 200 fighters. JG 27 successfully shepherded a single *Gruppe* of KG 30's Ju 88s to the outskirts of the city but the price was eight aircraft from the *Stab*, 1, 2, 3, 4 and 6 *Staffeln*. Of these, two 109s were destroyed and their pilots wounded in accidents, two were killed in combat and three were posted as missing to become PoWs. *Hptm* Eduard Neumann returned to Marquise to crash-land, writing off his aircraft in the process.

JG 51 lost three pilots killed while JG 52 had to record four Bf 109s 'failed to return'. In all these cases, the pilots survived and were captured. Among JG 53's losses this day was the aircraft flown by *Uffz* Poschenreider. He managed to put his Bf 109E-1 down near Rochester but had been seriously wounded in combat. Otherwise, this *Jagdgeschwader* lost three more Bf 109s, with one pilot being rescued from the Channel by the *Seenotdienst*. *Frei Jagd* sorties were flown throughout the day by JG 54, the unit recording two losses. One pilot was killed in combat and one became a PoW after putting his machine down near Bexhill, Sussex.

By the end of the day the *Jagdverbände* had lost 28 Bf 109s destroyed, with 26 pilots killed or missing. Seven more fighters required various degrees of repair and three wounded pilots would not be available for operations for some time to come. Even for a sizeable force, the loss of thirty-plus Bf 109s was at that time considered a terrible price to pay for one day's operations. The Bf 109 losses of 30 September were not exceeded during 1940 and the figure stood as a grim record of the attrition rate. It showed an alarming trend and one that *Luftwaffe* planners could hardly have foreseen occurring at the end of a battle they believed they were winning a few weeks previously.

With the bomber offensive now aimed primarily at London it was clear that the *Kanalkampf* had changed completely; given even a small respite, Fighter Command had clearly demonstrated its powers of recovery and at least to the Germans, to appear as strong in the air as it was when these operations began. Very little had been achieved by the *Luftwaffe* as a whole and the flyers realised the fact when on 12 October, Hitler indefinitely postponed Operation *Seelöwe* (Sea Lion).

When the loss/victory ratio achieved by both sides in the *Kanalkampf* was examined, it was not perhaps surprising that the Bf 109E had emerged with results as good as those of the Spitfire. Some of the figures placed the Bf 109 marginally ahead but overall the record of the opposing fighters was remarkably close. The Bf 109 of course was up against two enemy types and in those combats involving only Hurricanes, its pilots fared (again unsurprisingly) very well, for the record against the old Hawker fighter had always been good even though the *Jagdverbände* had by mid-1940 met more Hurricanes than any other enemy fighter type.

An unpalatable fact remained: Operation *Seelöwe* had not gone ahead because the *Luftwaffe* had been unable to establish air superiority over the landing beaches on the English coast. By giving the defences a further few weeks to prepare after the Dunkirk evacuation, the Germans lost the initiative and with it any last element of advantage that an invasion might then have exploited.

Viewed with hindsight, the *Luftwaffe*'s strategic task was all but impossible with the existing equipment and prevailing tactical doctrine little changed to adapt to a totally new kind of warfare; only the Bf 109 *Jagdverbände* had any real chance to succeed and the pilots did very well – as far as they were able with an aircraft that needed for one thing, a great deal more endurance. When they were optimistically tasked with destroying the RAF fighter force in an impossibly short time scale and then severely restricted by having to

The first Bf 109F-1s reached operational units by late 1940 and in early 1941, full production of the F-2 commenced. Some of the latter are pictured just after completion. *(Robertson)*

escort bombers, almost all their advantages were sacrificed. Few German pilots had at the outset of the battle realised the full significance of conducting the majority of their sorties and combats entirely over water, but once again, there had been no precedent.

Analysis of the relative merits of the Bf 109 and Spitfire had begun even before each fighter had met in combat and was to continue after what the British termed the Battle of Britain had been decided. On balance the two fighters were more or less equal in all the most important respects; there were those RAF pilots who swore that the Bf 109E was superior to the Spitfire Mk II, particularly in climb rate and speed at altitude – and some *Jagdflieger* who might have exchanged their *Emils* for Spitfires, had they been given the chance. Others who test flew the Supermarine fighter made much of the Merlin engine's propensity to cut out under negative G, little realising that this short-lived problem was not the Achilles heel it might have appeared to be.

Data on the performance of the principal fighter on each side extended to figures which revealed how well the Bf 109 had performed on those occasions when it was only in combat with Spitfires or Hurricanes. One assessment put the relative loss figures as: 219 Spitfires shot down for 180 Bf 109s and 272 Hurricanes for 153 Bf 109s, respective totals of 491 to 333.

A skilled pilot could get marginally more from his machine than the man of only 'average' rating and this often made the difference in combat. One thing was certain – the *Jagdflieger* appeared never to entirely forget the traumatic meeting with Spitfire squadrons over the English Channel. This fighter's outstanding manoeuvrability became legendary and despite all that came afterwards, those who had fought the RAF in 1940 – and even those who had not – retained a healthy respect for the first fighter to offer a real and ultimately successful challenge to their Messerschmitts.

Enter the F

Some of the first Bf 109F-1s to be built were delivered to *Stab./JG* 51 (HQ Flight of *JG* 51) on the Channel coast in October 1940. Having been built after a small number of Bf 109F-0s, the F-1 was also considered an interim variant, main production not getting underway until early in 1941 with the Bf 109F-2. Both the Bf 109F-0 and F-1 were powered by the 1,200 hp DB 610N, the engine intended for the series (the 1,300 hp DB 601E) not immediately being available. It was fitted in the F-2, however.

With only a handful of the new model available, the first priority was to convert the largest number of pilots, pending deliveries of enough examples to enable a full change-over from the *Emil* to be made as smoothly as possible. Few early Bf 109F models had been delivered to front line *Staffeln* before the tail weakness manifested itself and modifications further delayed complete re-equipment – but a few examples were flown in action in 1940 by experienced pilots, who also perhaps exercised the 'privilege of rank' in order to do so!

Jabos (Fighter bombers)

In late 1940 a new role for the Bf 109E furthered Göring's edict in October to the effect that fighter bomber attacks were to be made against England. This in turn was the result of favourable reports by pilots of 3./*Erpr.Gr* 210 and II(S)/*LG* 2, both of which had operated the BF 109E-1/B during the main daylight assault. In order to pool pilot experience and draw up a comprehensive set of tactics (such as was possible with single-seat fighters) for a highly exacting task, each *Jadgeschwader* based on the Channel coast formed a specialised fighter bomber *Staffel*.

Although individual units and pilots varied their tactics according to prevailing conditions, the generally accepted practice was to initiate a 15-20 degree dive using the *Revi* gunsight and point the nose of the aircraft about 200 yards ahead of the target. Airspeed built up rapidly as the dive angle increased until at 80 degrees, the IAS was up to 480 mph. Bomb release could not however be effected at so steep an angle for fear of throwing it forward into the propeller and the pilot had to pull back. After bomb release, the pull out was made at around 3,000 feet altitude. Saddled with the weight of the ETC 500 rack and a 551-lb SC 250 bomb, the speed of the Bf 109E dropped by some 30 mph, but the high-speed dive prior to release enabled the aircraft to show the defences a clean pair of heels and frequently evade interception.

Last of the *Emils*

During the spring of 1937, a number of Bf 109Bs were delivered to *Küstenjägergruppe* (Coastal fighter group) 136 of the *Kreigsmarine* for trials to determine the type's suitability for operation from Germany's first aircraft carrier, the *Graf Zeppelin*. This move, some 18 months prior to the carrier's launching, was one of the first practical steps towards forming an air group composed of ten Bf 109 fighters, 13 Ju 87C dive bombers and 20 Fi 156 observation aircraft. It was followed in February 1938 by the initial flight trials of the Bf 109 V-17 fitted with catapult gear and an arrester hook.

By the time the *Graf Zeppelin* was launched on 18 December 1938, a decision had been made to base the carrier-equipped Bf 109 on the E model, which thus became the Bf 109T i.e. *Trager* (carrier). Principal among the modifications was the addition of an arrester hook and catapult spools and extension of the wingspan from 9.85 metres (32 ft 4.5 in) to 11.08 m (36 ft 2.75 in) and incorporation of wing spoilers, although unlike on the Ju 87s intended for shipboard operation, no provision was made for wing-folding.

In 1938 a fighter unit, II./*Tragergruppe* 186, was raised and equipped with the Bf 109B to test the suitability of the *Luftwaffe* fighter for naval operations. The unit received some Bf 109Es in the summer of 1939 but was disbanded before the first Bf 109T flew.

Delays in completing the *Graf Zeppelin* led to a number of changes in the proposed production order for the Bf 109T but eventually it was decided that Fieseler at Kassel would build a batch of 70 aircraft, only seven being carrier-equipped T-1s. The balance of 63 were T-2 land-based machines with carrier equipment removed but incorporating the increased span wing.

These 63 unique Bf 109s were completed early in 1941 and as there was no immediate use for them, they were ferried to Denmark and subsequently to Norway by pilots of *JG* 77 to be placed in reserve. Of the total built, only 52 went abroad, as six were retained by Messerschmitt and two had been lost in accidents. On 28 September the Bf 109Ts were formally transferred from *Kreigsmarine* to *Luftwaffe* control and there followed a lengthy period of storage and regular overhaul before on 5 April 1943, the Bf 109T entered service.

In the meantime a new I./JG 77 was organised, separate to the original unit which was absorbed by JG 51. A *Gruppenstab* of the new I./JG 77 was reformed in January 1941 although a *Stabs-Staffel* had existed since July 1940 and had its origins in the *Stabs-Staffel* of JG 186. Bf 109Es formed the original equipment, I./JG 77 being commanded by *Hptm* Walter Grommes. The *Gruppe* had the standard three *Staffeln* based at Sola, Lister and Herdia respectively and Bf 109Ts were delivered in June 1941, 42 aircraft going to the three *Staffeln* and 21 to *Jagdgruppe* Drondheim.

JG 77 was tasked with coastal patrols and interception of the numerous RAF incursions (mostly by Coastal Command) over and around Norway and the long-span Bf 109Ts were first in action in June. British airmen failed to appreciate the subtle differences between the two versions and routinely reported being attacked by Bf 109E or Fs. The majority of RAF losses over Scandinavia were directly attributable to the Bf 109Ts and Es of JG 77, at modest cost to the Germans. The first T model to be lost occurred on a 19 June ferry flight.

There was now only a very slim chance that the Bf 109 would ever be flown from a carrier deck as work on the *Graf Zeppelin* had been suspended in 1940. Despite there being little chance of her being commissioned and actually undertaking war operations, the carrier was not broken up and some work on her was resumed during 1942-43. There was also discussion on building escort-type carriers for the navy, so the Bf 109T's original purpose was not entirely dead.

Chapter 5
Mediterranean Debut

With his sights set firmly on the conquest of the USSR, Adolf Hitler had made no plans for a wider war before that mighty undertaking had been completed and it was with some irritation that he noted the failure of his Italian ally to achieve much following an incursion into Egypt on 13 September 1940. Benito Mussolini's annexation of Albania was to be followed by the securing of large tracts of North Africa as a basis for a new Roman empire extending throughout the Mediterranean, a plan that quickly began to founder in the face of spirited British resistance.

In the Mediterranean the strategic importance of Malta was highlighted by repeated attempts by the *Regia Aeronautica* to neutralise the island's airfields and port facilities and thus deny Allied domination of the sea lanes – vital to supply any force undertaking a military campaign in North Africa. This it singularly failed to achieve.

British determination not to give ground in Egypt and thereby concede Axis domination of the entire region including the Arabian oil fields and the Suez Canal, led to the despatch of reinforcements, modest though these were until the threat of invasion of the British Isles had been contained. And although outnumbered manifold the British Army held its ground in a

When the *Jagdwaffe* began operations in North Africa, JG 27 personnel found spartan surroundings and a lack of kit. White theatre bands adorn all the I.*Gruppe* Bf 109E-4s visible.

Typical desert landing grounds used by *JG* 27 in Libya featured open air dispersals for the Bf 109s and few buildings (one being visible here), but acres of flat ground were an asset.

number of desert duals. It was shortly before the Allies began a gradual advance against the Italians on 7 December 1940 that the Germans decided, not entirely to Mussolini's satisfaction, to intervene.

Initially, no *Luftwaffe* single-seat fighters were sent to the Mediterranean, although bomber and *Stuka Gruppen* were supported by Bf 110 *Zerstörers*; Fliegerkorps X was transferred to conduct an anti-shipping campaign, to generally harry the British and assist the Italians – which included supporting the *Regia Aeronautica*'s assault on Malta. The German units had been established in Sicily by December 1940 and in February 1941 the first Bf 109Es of *Oblt* Joachim Muncheberg's 7./JG 26 arrived from France to make life less hazardous for bomber crews, whose aircraft were invariably intercepted by Malta's small force of Gladiators and Hurricanes.

These particular Bf 109Es were the 'advance guard', for Muncheberg's *Staffel* was not destined to stay permanently in the Mediterranean. Apart from briefly withdrawing from Sicily to fly sorties over the Balkans in April, the dozen or so Bf 109Es of 7./ JG 26 concentrated almost totally on wearing down the defences of Malta, a task the handful of pilots carried out very efficiently for nearly seven months. Muncheberg, a *Ritterkreuztrager mit Eisenlaub*, was then one of the *Luftwaffe*'s most successful fighter pilots with 43 victories to his credit.

Fighters however, are unable to decide strategic campaigns; only by completely destroying Malta by bombing or capturing it by seaborne or glider assault, could the Axis finally remove the island as a threat. The latter option did not appeal to Hitler while the defences remained active. Equally, he felt unable to abandon his Axis partner and ordered the expansion of the *Luftwaffe* 'expeditionary force' concurrent with the arrival of the *Afrika Korps* under Erwin Rommel.

While Muncheberg's small force was making

MEDITERRANEAN DEBUT

life unpleasant for the Malta defenders, the Western Desert witnessed the arrival of the first Bf 109Es, in the hands of I./JG 27. Despite the late 1940 debut of the Bf 109F in Europe, this model was not immediately sent to the area for the *Emil* remained an excellent aircraft, well able to contest skies ostensibly dominated by a motley collection of enemy aircraft, many of which had seen better days. Based at El Gazala in Libya, *Hptm* Eduard Neumann's *Gruppe* flew its first combat sorties on 19 February; JG 27 rapidly made a name for itself, not only for its prowess in aerial combat but sound battle tactics.

Unlike their adversaries, the German *Jagdfliegern* rarely departed from their proven 'finger four' formation, the ethos of mutual wingman support in a dogfight and the maintenance of a high cover to formations operating at lower altitudes. Good use was also made of the angle of the sun, the *Jagdfliegern* invariably placing themselves between the dazzling light and the target, forcing enemy flyers to stare upwards blindly as the Messerschmitts fell on them. Many of the I *Gruppe* pilots had seen action and more than a few had scored aerial victories. These were soon added to as the peculiar nature of the North African ground war dictated the degree of air activity over the various, quite localised front line positions.

In the desert, ground forces held the keys to ultimate victory; it was their task to hold seaports for the unloading of supplies, secure the roads (such as they were) and defend a network of airfields. Fighting tended to be limited to the coastal strip of North Africa where in particular the ports of Tripoli and Benghazi took on major significance.

Apart from the primary purpose of providing pin-point bombing attacks designed to destroy selected targets in concert with the *Afrika Korps*,

Badly bent Bf 109E-4/B of ZG 1 down in the Western Desert.

'Black 2' of I./JG 27 carrying a long range tank, reaches the point of no return during combat operations in North Africa.

Luftwaffe air support in the desert was varied and often indirect, ranging from attacks on individual columns of troops, tanks and vehicles and aircraft to a sustained assault designed to pave the way for the army during major offensives such as the capture of Tobruk. For the fighter *Gruppen*, every enemy ship, tank, truck or aircraft destroyed or disabled, every fighter, bomber or transport destroyed in aerial combat, added to the attrition rate that had to be borne by the enemy.

Once more the *Luftwaffe* enjoyed excellent support from the men who flew the ubiquitous Ju 52, a service doubly appreciated in the desert where the barest minimum of facilities existed to support fighter operations. As the Germans had made few plans to fight a war under desert conditions, *Jagdwaffe* personnel found all types of equipment and vehicles in short supply. This included clothing more suited to the hot/cold extremes of desert days and nights and many such items had to be purchased locally.

Fortunately, the needs of the fighter *Staffeln* were fairly basic; providing that they had fuel, oil and ammunition, all of which was stockpiled in ample quantities, both on the airfields and at depots, they could operate on a daily basis and fly multiple sorties in the course of a single morning or afternoon as necessity demanded.

Flying had occasionally to be cancelled or curtailed due to prevailing weather conditions as the desert climate, while offering clear air and unlimited visibility in calm conditions, could also bring sand storms and torrential rain, which could rapidly turn desert airfields into quagmires. At such times, units were obliged to change base or await dryer conditions before continuing operations.

In general the *Luftwaffe* engineering sections coped with the constant high temperatures – often reaching 120° C – and the perpetual presence of fine sand. The Bf 109E's Daimler-Benz DB 601 engine stood up surprisingly well to a good deal of open-air servicing under primitive conditions. Bf 109Es earmarked for operations under desert or 'tropical' conditions had been fitted with filters over the supercharger air intake to prevent engine overheating or the ingestion of abrasive sand particles that could cause extensive damage.

German pilots serving in the desert also found an almost unique freedom. A virtual absence of strategic targets, few civilian dwellings, the nature of the terrain and the fluctuating, relatively small scale ground war did not restrict them to a rigid pattern of close support and left them able to indulge in numerous *frei Jagd* sorties which were as damaging to the enemy as any other form of operation, probably more so. They did however, have the routine chore of escorting *Gruppen* of Ju 87s and Ju 88s – which was one of their primary reasons for being in the Mediterranean.

With help from the *Regia Aeronautica* fighter *Squadriglia*, JG 27 flew numerous escort sorties to Ju 87s and Ju 88s and attempted to protect their charges from interception by Allied fighters.

Hans-Joachim Marseille gets a helping hand with the harness of Bf 109F 'Yellow 14', just one of a number of similarly-marked aircraft he flew in III./JG 27.

With each side flying much the same type of aircraft that had previously been met in combat over Europe, relative performance characteristics were familiar to any pilot who had served in his respective unit even for a brief period – in short there were few surprises in store for the *Jagdwaffe* or the RAF in terms of the aircraft types deployed by the other side.

What did surprise the Germans was the apparent lack of air discipline and ground direction of the British fighter squadrons compared with that

As the desert war waxed and waned for the Germans, the *Luftwaffe* was forced to abandon scores of repairable Bf 109s including this *Friedrich*, property of JG 77. (*via Marsh*)

Wrecked Bf 109s and Ju 87 fuselages, the debris of the desert war, include Bf 109F 'yellow 12' of I./JG 3 among the variety of paint schemes.

displayed to such advantage on the Channel front. Aircraft were often observed flying in unwieldy flocks with little evidence of top cover – or indeed of much experienced leadership. For many *Jagdflieger*, the desert became a far easier arena in which to score victories.

While some early Western Desert fighter operations left much to be desired on the part of the RAF, this was due in part to the rapid expansion after the end of the Battle of Britain. Many seasoned pilots were posted overseas where they were invariably considered more valuable occupying command positions on the ground rather than the cockpit of a fighter. In many cases, it was the younger and greener newcomers who tended to bear the brunt of the fighting in the early months of the *Luftwaffe*'s tenure in the desert.

For their part, RAF pilots, however inexperienced, could hardly have failed to have heard about the poor showing of the Ju 87 against Hurricanes and Spitfires defending England. And here they were, faced with that same adversary, one which the most inexperienced fighter pilot should have little difficulty in shooting down. Consequently, the *Jagdwaffe* was saddled with a doubly difficult escort task. With enemy pilots knowing full well how easily they could destroy *Stukas* if they could avoid the Messerschmitt fighter screen, they had their work cut out to protect the dive bomber crews. It was never an easy task due mainly to the widely differing performance of each aircraft, but for some time the Ju 87 remained one of the *Luftwaffe*'s primary weapons in the theatre.

Fighters evened the odds by indulging in ground strafing of the opposing side's airfields, a practice that while hazardous, could reap great benefit. It was hard to camouflage aircraft and vehicles adequately against the largely dead flat, featureless terrain and anything that looked to be of use to the enemy was fair game. The Western Desert was a war theatre where fighter bombers came into their own, originally as an operational necessity to fill a gap created by a general lack of conventional bombers.

Despite the fact that *JG* 27's Bf 109Es were equipped to carry bombs or drop tanks, the targets often lay well within standard flying range and pilots tended to fly their aircraft in lightweight configuration and to favour gunnery attacks on ground targets whenever possible. A specialist *Jabo Staffel Afrika* equipped with the Bf 109E was subsequently formed to undertake fighter bomber sorties almost exclusively.

The many deprivations of the desert seemed to have little adverse effect on the eager young *Jagdfliegern* and the pilots and groundcrews of *JG* 27 soon came to terms with their new surroundings in Libya. Those pilots who had arrived in the desert with modest aerial victory scores were quickly – and sometimes surprisingly easily – able to increase their tallies of enemy aircraft destroyed. Overall, I./JG 27 did well without sustaining high casualties, for many months.

Heading the list of top-scoring pilots was *Oblt* Gerhard Homuth with 15 victories, followed by *Oblt* Ludwig Franzisket with 14. *Oblt* Karl-Heinz Redlich had 10 and seven enemy aircraft had fallen to *Lt* Willi Kothmann. I./JG 27's pilots had arrived in the desert with a combined victory total of 61 and almost immediately the *Gruppe* commenced operations in Libya it became clear that the Bf 109E had found a new arena to extend its enviable combat record. Pilots often flew in no more than *Schwarm* or *Rotte* strength, yet by stressing the importance of maintaining a high cover and instilling in every man the necessity of constantly scanning the sky, they were rarely surprised by enemy fighters.

Target aircraft for the *Jagdflieger* during this early period were many and varied, with an equally wide range of performance characteristics and armament. But as always, it was air discipline, planning and skill that made the difference in air combat. Most importantly, the desert was a theatre where novice pilots had time to observe and emulate the *Experten* whose shoes they would one day fill. The system of mutual protection by the leader/wingman helped foster a high degree of responsibility and general awareness of the prevailing tactical situation.

On an average daily sortie the *Luftwaffe* tactical planners would position fighters at the relatively low altitude of 6,000 feet, this being found to be favourable to the Germans in most situations. Sorties were directed by separate German and Italian command posts although the *Regia Aeronautica* often provided fighter escort to German dive and conventional bombers, a sound move as it freed the Bf 109s for their invariably successful *frei Jagd* sorties which resulted in a high number of air combat victories.

Fighter patrols were mounted throughout the daylight hours, from dawn to dusk. A considerable amount of combat took place at the end of the day when the low angle of the sun offered pilots some protection from the intense glare. This time of day was particularly welcomed by Allied aircrew, particularly by those flying the relative slow types which were vulnerable to fighter attack, such as the Blenheim, Maryland and Wellington.

Yugoslavia and Greece

Having persuaded Bulgaria to join the Axis Pact, Hitler was preoccupied with the forthcoming invasion of the Soviet Union when on 26 March 1941, Yugoslavia tore up an agreement to side with Germany. Fearing that an Allied presence in that country or Greece could endanger his attack on Russia, Hitler then moved to secure the Balkans. Military operations against Yugoslavia began on 6 April.

The Bf 109 units (all of them flying the E model) assigned to the invasion of the Balkans were: II./JG 54 and I./JG 27 under *Fliegerführer Graz* which otherwise had a *Stab* and one *Gruppe* of

Ju 87s; *Fliegerführer* Arad in Romania contained III./*JG* 54, plus 4. *Staffel* and *Stab*, II and III./*JG* 77. Based in Hungary but available to *Fliegerkorps VIII* were *Stab*, II and III./*JG* 27 and I(*J*)LG 2. The force ranged against the ill-equipped Yugoslavs was potentially overwhelming as *Regia Aeronautica* air units backed up the *Luftwaffe*.

Fighter versus fighter combat over Yugoslavia was understandably brief and occasionally very confusing, as the *Luftwaffe* Bf 109Es came up against the *Emils* supplied to the *JKRV* before the war. The latter force had just over 60 Bf 109Es in service at the time of the German attack but once again, a nation faced with *Blitzkrieg* tactics was given precious little time to react effectively. Many aircraft were destroyed on the ground during the initial phases and poor communcations did nothing but hinder a confused situation. It was a manifestation of an unpreparedness for war that had been fatal elsewhere in Europe and the citizens of Belgrade suffered the grimly familiar experience whereby a German air attack on the capital city all but sealed the fate of a country. Yugoslavia surrendered on 17 April.

Concurrent with the attack on Yugoslavia, German troops invaded Greece, a rapid ground advance being made under the *Luftwaffe's* efficient umbrella, which wreaked destruction on strong points, supplies and airfields. In combat the small RAF contingent assisting the Greeks with Hurricanes, Blenheims and Gladiators was almost powerless to prevent precious (but generally obsolete) aircraft being destroyed during widespread strafing and bombing of the airfields. Few machines remained operational after 21 April. An evacuation saved as many Allied troops as possible but the Greek government had little choice but to surrender on 27 April.

During April 1941, the *Luftwaffe* lost 182 aircraft, the majority of them during war operations in the Balkans. The statistics of the campaign included 54 Bf 109Es lost against claims of 110 Yugoslavian, Greek and British aircraft shot down by Bf 109s and Bf 110s.

The successful German operation in the Balkans was reflected in Axis gains in North Africa and by April 1941 the *Afrika Korps* had Tobruk under siege. *Luftwaffe* support for Rommel was excellent although the Bf 109 pilots once again had their work cut out to cover the *Stukas* which could fall victim to groundfire once their high speed dives had been completed. A Ju 87 pulling out of its dive and throttled right back to the point where it could be picked off even by a well-aimed Bren gun, was an image that haunted many a Messerschmitt pilot!

Nevertheless the Germans succeeded in achieving and maintaining local air superiority, I./*JG* 27 not being exactly over-stretched when Allied fighters came on the scene. With the RAF still apparently believing that 'V' formations, complete with 'weavers' and a 'tail-end Charlie' to guard the main body were tactically sound, the seasoned *Jagdflieger* could hardly help but score victories.

Also, the outmoded defensive circle continued to be employed by the enemy. This, designed to provide mutual protection, exploited the good turning circle of the Hurricane and could be all but impenetrable if maintained. But it remained defensive and therefore hardly productive in terms of aggressive combat tactics; some of the bolder German pilots risked their necks to break into the circle and pick off victims – but taking on enemy formations that outnumbered them many times over was a regular occurrence for the desert *Jagdflieger*.

Not that the RAF failed to realise the disadvantages of its *ad hoc* tactics and to amend operational directives. The German flyers observed ruefully that although the enemy might not have been able to teach them much about tactics, he was invariably present in gradually increasing numbers. The variety of types the British and Commonwealth squadrons deployed was perhaps surprising, although bombers such as the Blenheim, Beaufort and Wellington had never represented much of a challenge to a well-flown Bf 109. It was however foolhardy to believe that things would not change.

While *JG* 27 continued to be successful, the ground war had deteriorated into a stalemate when Rommel was forced to abandon his attempt to take Tobruk to await reinforcements and supplies. The German fighter force alone could do nothing to alter that situation and in general, the *Luftwaffe* in North Africa was never able to achieve the kind of air/ground co-operation that the Allies were gradually building. Rommel's command, subordinated as it was to the Italians, was never over-endowed with German ground divisions or aircraft, a situation that hardly helped the tactical planning for a successful campaign in a difficult theatre, one regarded by Hitler as a mere 'sideshow' to the main business of subjugating the Soviet Union.

In May 1941 the British decided to hold Crete for as long as possible, the island being reinforced by troops evacuated from Greece. This move brought almost immediate German response in the form of a paratroop assault. Charged with covering the *Luftwaffe* transports carrying men of the *Fallschirmjäger* to Crete from Sicily was a combined force of Bf 109Es of II. and III./JG 77 and I.(J)/LG 2. By the time the assault began on 14 May the RAF presence on the island had been drastically reduced in combat with Bf 109s – but the troops had had time to make excellent use of all available cover. Low flying German fighters found ground fire extremely hazardous, more so than enemy fighters. Gradually the Germans gained the upper hand, despite heavy casualties and the Royal Navy was alerted to evacuate survivors of the bitter fighting. But to achieve that undertaking, the British ships had to sail well within range of a powerful *Luftwaffe* striking force in Sicily.

On 22 May, the two *Jabo Staffeln* (5. and 7) of JG 77 commanded by Lt Huy and II.(S)/LG 2 under Herbert Ihlefeld, attacked British ships off Crete. Then based on Molaoi on the Elos peninsula of the Polopouresi, the fighter bombers, in company with *Stukas* and medium bombers, decimated the cruiser and destroyer forces engaged in the evacuation of troops from Crete.

A *Rotte* of Bf 109Es of LG 2 was subsequently credited with causing fatal damage to the cruiser *Fiji* by placing their 500-lb bombs accurately enough to blow in the ship's bottom. She rolled over and sank, apparently after being hit by three more bombs from a solitary medium bomber. Having also scored hits on the battleship *Warspite* on the 22nd, the action continued for the *Jabo* specialists on the following day, in the course of which Maj von Winterfeld, leading elements of III./JG 77, caused havoc among the MTBs of the 10th Flotilla by sinking five boats.

By capturing Crete, Hitler's timetable for securing a southern flank in the Balkans was completed without the planned invasion of the Soviet Union in July having to be postponed for more than a few weeks. But the battle had wiped out the cream of the *Fallschirmjäger*, making a similar airborne assault on Malta a virtual impossibility. It was early June before all resistance ended on Crete, by which time JG 77, LG 2 and III./JG 52, the latter a reinforcement called upon but not required in the fight for the island, had received new orders to move to eastern Europe to become part of the massive *Luftwaffe* task force for Operation *Barbarossa*.

For III./JG 52, the relocation to the East marked the first exposure to combat since the summer of 1940. A *Gruppe* that had suffered particularly high losses during the *Kanalkampf*, it had spent nearly a year in Germany in reserve and was now primed to re-enter the fight. In the interim, 8. *Staffel*'s Bf 109s commanded by Gunther Rall, had been stationed in Rumania to help defend the Ploesti oil refineries and to provide the cadre, along with pilots of 9. *Staffel*, of I *Gruppe* of a new *Jagdgeschwader*, JG 28. In the event JG 28 which had existed for barely six months, was reabsorbed by JG 52. By the time of *Barbarossa*, 9. *Staffel* had passed to the command of Hermann Graf.

I./JG 27 remained in Libya while II and III *Gruppen* had also in the meantime, been alerted for duty in the desert. They transferred respectively in September and December 1941, II *Gruppe* having by then spent some three months in Russia. In the meantime, the nomadic 7./JG 26 had become more than familiar with Gela, Sicily and had briefly served alongside I./JG 27 before returning to France in August. During its time overseas, the 'Red Heart *Staffel*' had been outstandingly successful and had scored 52 kills without losing a single aircraft in combat.

The already diverse list of Allied aircraft opposing the *Jagdfliegern* over the desert was increased further in June 1941 when the first P-40B/C Warhawks and P-40D/E Kittyhawks operated by RAF squadrons, were encountered. Neither American fighter, though well armed and armoured, gave the desert *Experten* too much trouble – in fact their victory claims and combat reports began to include the word 'Curtiss' (the Germans' universal generic term irrespective of the correct sub-type) with increasing regularity!

Things on the RAF side slowly improved however and the Germans' understandable failure to distinguish readily the finer points of early and later model P-40s could work against them. The P-40E for example had six of the heavy .50-in machine guns against the .303s installed in the P-40B and as pilot experience built up, so the Allies could count much more on the Kittyhawk, even in a dogfight with Bf 109s. But the Bf 109 itself was about to be improved as far as the desert *Jagdfliegern* were concerned. The lull in the ground war that extended into the autumn of 1941 allowed I./JG 27 to despatch a *Staffel* at a time back to Germany to re-equip with the Bf 109F-2/Trop.

While the Bf 109F was modestly armed for the standards of the day its centrally-mounted armament eliminated the problem of harmonisation common with wing-mounted guns. It meant in effect that the pilot merely had to close the range and point the nose of his aircraft, almost guaranteeing hits if he had calculated the deflection correctly. That said, pilots still needed a good 'eye' in order to score victories and those whose skills were only 'average' really needed the better spread of fire that fuselage and wing guns produced. As in all war theatres, the Germans found that the bulk of air victories fell to a relatively small number of pilots.

Any problem with firepower that could be a little too concentrated for some pilots was subsequently overcome by additional guns attached to the wing underside outboard of the wheel wells. Widely used on the Bf 109G, these gondola-mounted MG 151/15 cannon first became available for the F-4 as *Rüstzustand* 1 (Conversion set). A total of 250 F-4/R1 aircraft were built and among the units to actually deploy the aircraft with the additional cannon were JG 3 and JG 52, while JG 77 is known to have also used some Bf 109F-4/R1s, which lacked the necessary wiring for the gondola-mounted cannon.

Crusader

By opening an offensive code-named Operation CRUSADER in November 1941, the Allies made some substantial gains, much to the confusion and surprise of the Germans. Forced onto the defensive, the *Afrika Korps* gave ground, with the *Luftwaffe* temporarily bogged down on airfields soaked by recent rains. Rommel recovered quickly however and the *Luftwaffe* fighter force, repositioned on drier airfields, reacted vigorously. Gazala, JG 27's main base since arrival in the desert, was abandoned in favour of Martuba for the time being. A welcome event was the arrival of II and III *Gruppen* to bring the *Geschwader* together for the first time since 1940.

This phase saw numerous strafing sorties against airfields by both sides in an attempt to reduce air attacks on the ground forces and *Luftwaffe* operations began to slacken, not only as a result of aircraft being destroyed on the ground, but low fuel supplies. On 20 November I./JG 27 shot down four South African Air Force (SAAF) Marylands to thoroughly demoralise the surviving crews and the BF 109s became involved in a big air battle with RAF fighters on the 22nd. Despite flying the Bf 109F, the Germans found the Tomahawk and Hurricanes more than a handful on this occasion and lost six of their own for ten fighters and four Blenheims.

This might have appeared a good ratio for the Germans, but the loss of six aircraft in one day was serious for the small force the *Jagdwaffe* always represented in comparison with the enemy. One should not forget the contribution made by the *Regia Aeronautica* but this too suffered from finite numbers of fighters and to an extent, the independent operations mounted by both Axis air forces. Co-operation was often achieved but large scale, fully integrated fighter sweeps for example, were rarely flown.

On 6 December the Bf 109Fs of III./JG 53 arrived in the desert from Sicily to occupy Derna, a base it had to evacuate shortly before Christmas. The fuel shortage was partially met by flying supplies in from Crete in Ju 52s. Always vulnerable, the transports were attacked by the RAF and some were lost. Four days later the Bf 109s shot down two examples of a type new to the theatre, the Douglas Boston. Such success in the air was offset by the dire situation on the ground; by pulling back, the *Jagdwaffe* had to abandon a great many fighters under repair. To make matters worse, there were losses of pilots in combat, among them III./JG 27's *Oblt* Erbo Graf von Kageneck, mortally wounded in a dogfight with Hurricanes on 24 December. With 67 victories von Kageneck was one of the *Luftwaffe*'s top *Experten*. He died of his wounds on 12 January 1942.

With supply lines over-stretched, a situation hardly helped by the pressure being put on Malta by Axis air attacks, CRUSADER ran out of momentum and by February 1941, the Germans had consolidated their position at El Aghelia. Allied reinforcements to the desert then suffered through a bleed-off of troops and supplies to the Far East to counter the Japanese and few further gains could be made.

The renewed *Blitz* on Malta achieved – albeit temporarily – the desired effect of putting Allied fighters on the defensive and keeping bomber sorties down and Rommel, his tanks able to roll again, began to advance. Most of the former *Luftwaffe* airfields were back in German hands by January 1942.

With no shortage of targets, the *Jagdflieger* continued to make life difficult for Allied fliers. Hans-Joachim Marseille brought his total of victories to 40 on 9 February and Homuth was catching up rapidly – as was Otto Schulz of II *Gruppe*, who had 51 by the time he left the theatre later that month.

Malta was still under virtual Axis blockade and despite the first Spitfires arriving on the island on 7 March, the Allies could make little headway other than offering a consistently brave defence which continued to take a toll of German and Italian bombers. Fighter combat over the island invariably saw the boot on the other foot as the *Jagdfliegern* were not unduly worried by the presence of Hurricanes, even in greater numbers.

Neither did the newly-arrived Spitfires, the supposed master of the Bf 109, give much trouble. Pitted against the Mk V, the Bf 109F often came off best, particularly as at that stage, the RAF pilots were not generally very experienced.

Germany was however, never able to fully exploit any gains in the Mediterranean after the start of the campaign in the East. No sooner had Malta been contained, thus allowing Rommel to advance with only localised enemy air attacks to deal with, than there was a new development in Russia which demanded transfer of *Luftwaffe* forces. Such occurred early in May and among the departees were the Bf 109s of II./JG 3 which had been in the theatre for only a few weeks and I./JG 53. Up against largely Italian forces, the Malta defenders were able, as they had in the past, to be much more effective.

Rommel had meanwhile recaptured Benghazi and was pressing British defences at Gazala. *Fliegerführer Afrika* (Commander Air Forces Africa) determined to offset the advantage of Allied reinforcements, brought II./JG 53 to Gazala from Sicily to join his dive bomber and *Zerstörer* force. Italian units were also strengthened at that point and on 26 May when Rommel made his move, the Axis air forces gave staunch support. Between 29 and 31 May, the RAF lost 39 fighters in action.

As efficient as they were in downing enemy fighters, the *Jagdfliegern* very rarely managed to penetrate the escort and shoot down British bombers. The latter consequently took an increasingly heavy toll of *Afrika Korps* supply lines and forward positions. The other side of the coin was that the Ju 87s became less and less effective if intercepted by enemy fighters.

Having broken the Allied defences at Gazala and recaptured Tobruk, the Germans pressed on towards Rommel's ultimate goal, the Suez Canal. This ambitious plan had first to take prepared 8th Army positions at El Alamein, located between the sea and the impassable salt marshes of the Qattara Depression.

In the air, the first Mk V Spitfires met in combat by the desert *Jagdflieger*, did not fare as well as the British had hoped. Rommel, having been repulsed in his first attacks at Alamein, marked time. Long supply lines, Allied pressure and the sheer exhaustion of his troops forced him to regroup and to lay plans for a new move, against what he believed to be weakly-held positions on the edge of the Qattara Depression, at Adem El Halfa. Now commanded by Bernard

Montgomery, the 8th Army had been revitalised and reinforced to such an extent that any further setbacks were much more unlikely. Anticipating the German plan, Montgomery had reinforced the positions at Adem El Halfa.

On 31 August the German attack began. Overhead the *Luftwaffe* fighters took on their RAF opposite numbers which were in turn, out to decimate the *Stuka* formations supporting the *Panzers*. Marseille, back in the desert after a spell of leave in Germany, claimed 17 fighters shot down on 1 September, this out of a total of 20 lost overall.

Frei Jagd patrols over the lines invariably brought success to the Germans in combat and on this day, Marseille was briefed to escort Ju 87s, the Bf 109s taking off at 0730. Setting an easterly course they rendezvoused with the *Stukas* ten minutes later. Climbing to 11,500 feet, Marseille and his pilots soon spotted a formation of ten P-40 Tomahawks and dived to attack.

The RAF pilots appeared not to have noticed the imminent danger and stayed in formation as Marseille came within firing range and closed to 300 feet of his intended target. A short burst of fire shattered the Tomahawk's canopy, killing the pilot. It was 0820. Barely noticing the stricken fighter falling away, Marseille shifted his aim to his victim's wingman. Another short burst and another pilot was dead in a shattered cockpit. The RAF machine plunged down to crash less than a mile from Marseille's first victim. A minute had passed since the first victory.

The respective formations broke into a series of skirmishes. At 0830 Marseille latched onto a Tomahawk shadowing a Ju 87. Intent on attacking the dive bomber, the Allied pilot suddenly saw his wing root and cowling dissolve under a

Photographed in Romania, the yellow fuselage theatre identification bands of these III *Gruppe* Bf 109E-3s indicate imminent deployment against Russia. (*Holmes*)

stream of fire. At 0833 Marseille's third victim crashed in the desert.

Then six Spitfires joined the fray, firing as they dived on the Bf 109s. Marseille remained calm, judging deflection perfectly. He waited for the leading fighter to line up on his tail and close to less than 400 yards. Marseille executed a violent break and all six Spitfires overshot.

Marseille rolled, opened his throttle and closed up behind the British machines, now turning to re-engage. For one the move was too late. Using his cannon, Marseille created a fourth smoke column on the desert floor. It was 0839.

Low on fuel, the Germans broke off and returned to base, the last fighter touching down at 0914. Marseille's combat had left little work for the armourers as his cannon had used just 20 rounds and his machine guns, 60.

A second late morning sortie and a third in the afternoon brought Marseille more success until by 1753, his score had reached 17 for the day. This outstanding feat of arms brought Marseille further accolades and he rampaged through to the penultimate day of September to the tune of 44 more aerial kills, bringing his total to 158, all but seven of them in Africa.

Since June 1942 German fighter units had been re-equipping with the Bf 109G-2 but it was approaching autumn before I./JG 27 was able to replace its *Friedrichs*. On 30 September Marseille flew one of the first G-2s delivered to the *Gruppe* on a patrol. En route back to base, the desert *Experten* noticed smoke seeping into the cockpit. Trying in vain to determine the cause of the fire, he prepared to bale out if necessary. With the engine fire getting worse, he had little choice but to jump – only to suffer fatal injuries when his body smashed into the tailplane.

The loss of Marseille was a terrible blow to JG 27 for he was an unforgettable character who was sorely missed. Other pilots were however maintaining the *Geschwader*'s tally of kills, among them Gerhard Homuth and Hans-Arnold Stahlschmidt. On the debit side, II./JG 27 numbered among its casualties *Gruppenkommandeur* Wolfgang Lippert and 14-victory *Experte Oberfeldwebel (Obfw)* Albert Espenlaub of I *Gruppe*. Shot down and captured on 13 December 1942, Espenlaub unwisely made a bid for freedom after being captured and was promptly shot dead.

The Axis blockade of Malta was not to last and the *Afrika Korps* was obliged to fight the 'decisive' but ultimately disastrous battle of El Alamein in November. This defeat all but destroyed Germany's hopes of victory in North Africa. Also, by the late summer of 1942 plans had been finalised for the amphibious landing of a substantial Allied force in Algeria and French Morocco which would eventually trap Rommel between two armies.

Despite being consistently successful in air combat, the *Jagdwaffe* lacked the resources to do more to turn the Allied tide and could only attempt to reduce the effectiveness of Allied air raids. These grew increasingly heavy and instrumental in destroying German aircraft as airfields were systematically pounded by medium and heavy bombers. Some reinforcements arrived in December 1942, including II./JG 26 from France. III./JG 27 remained on Crete for the time being but the Operation TORCH landings would bring further action soon enough.

Torch

News that the Allies had succeeded in putting troops ashore at Algiers and Oran on 8 November 1942 meant a re-positioning of now-meagre fighter forces; fortunately for the Germans, many of the Allied air units supporting TORCH were new to combat, their pilots not always finding the well-seasoned Messerschmitt *Jagdflieger* easy to counter effectively. Despite their modest numbers, the *Jagdgeschwadern* remained a dangerous threat to Allied air operations during the initial phases of TORCH, which saw the Bf 109s in combat not only with the more familiar Allied types, but naval F4F Wildcat fighters and gunnery spotting SOC-3 Seagull floatplanes operating from capital ships. Bf 109s shot down a number of the biplanes, which had the unenviable but vital task of directing cruiser fire onto German strongpoints and tanks above contested areas.

But as before, the German fighters' efforts could only be partially effective; they were no longer a support force for a strong and agressive ground army and Allied strength – and tactical acumen – could only improve. Nevertheless, the theatre was by no means abandoned and the *Jagdwaffe* began a series of base, unit and personnel changes in order to keep fighting with ever-dwindling resources that well earned it the nickname of the '*Luftwaffe*'s fire brigade'. This epithet was not the sole prerogative of the fighter force as other elements of the *Luftwaffe*, particularly the dive bomber *Gruppen*, felt they were increasingly asked to quell enemy activity all over the front at short notice. The term was in fact used in all war theatres.

In the desert the Bf 109 remained the primary German fighter type; there were some small-scale transfers of Fw 190 *Jagdgruppen*, but the main effort by this type in the Mediterranean was in the hands of *Schlagtgruppen* (fighter ground attack). Even this was on a relatively small scale, not great enough to have much bearing on events.

An indication that the *Jagdwaffe* would more than have its hands full after TORCH came on 16 November when B-17s of the 97th Bomb Group carried out their first raid in the Mediterranean area by attacking Sidi Ahmen aerodrome at Bizerte. The Fortresses soon had a fighter escort in the shape of the P-38 Lightning, a type that the Bf 109 *Jagdfliegern* were often able to dogfight successfully. Spitfires in American hands proved more of a challenge and on the 30th, the 52nd FG claimed a single Bf 109G to mark its first victory in the theatre. With radar early warning of incoming raids, the *Luftwaffe* was able to position fighters advantageously and on 3 December the Bf 109s shot down three and possibly two more P-38s – but three Messerschmitts also succumbed during the resulting air action.

Bf 109 strength in the Mediterranean at the end of November 1942 was 172 aircraft, plus 12 tactical reconnaissance fighters. Ninety-five fighters were available in Tunisia and 15 in Sicily. These were not of course the only offensive machines the Germans boasted – but they were the most effective. By comparison, the Allies could base almost as many aircraft as the entire German fighter force on half a dozen airfields.

As well as the heavies, the Germans were faced with US light and medium bombers, B-25s, B-26s and A-20 attack bombers, all in burgeoning strength, and all of them with fighter escort, primarily from P-38s. Lightnings flying reconnaissance and ground-strafing missions also proved dangerous, particularly when the latter had fighter airfields as their targets. The *Jagdfliegern* were additionally up against P-40s and P-39 Airacobras flown by American and French pilots.

These newcomers ably supported the RAF veterans of the Desert Air Force equipped with Spitfires, Kittyhawks, Hurricanes, Beaufighters, Bostons and less able types such as the Bisley, Maryland and Blenheim. And however easy it may have been for the German *Experten* to shoot down even the best of these types on an individual basis, they were increasingly outnumbered to the extent that individual prowess became swamped. All enemy aircraft were now likely to be seen in increasing numbers – after TORCH the desert war totally changed to become even more of a battle of logistics and production.

In an effort to neutralise part of Allied air strength on the ground, the Bf 109 *Gruppen* carried out a number of raids on Thelepte and other airfields in Tunisia during January 1943, escorting both Ju 88s and their own fighters by flying a top cover while other Bf 109s made strafing runs. Rommel had meanwhile made some progress against green US troops by inflicting a humiliating defeat at Kasserine but he was no longer in a position to fully exploit gains, having to contend with two powerful Allied armies, the primarily American force to the west and Montgomery's 8th Army to the east.

For the *Afrika Korps*, the beginning of the end was not long in coming. By 4 April, after much hard fighting, the TORCH force and the 8th Army linked up, trapping thousands of German troops in Tunisia. Three massive rescue attempts by the *Luftwaffe* transport force saw terrible casualties inflicted by Allied fighters, nearly 100 aircraft being shot down. German fighters offered what cover they could, but guarding slow transports against overwhelming numbers of Allied fighters was all but impossible to do effectively.

Events now moved swiftly. Tunis and Bizerte fell to the Allies on 7 May and Axis forces were prevented from utilising the natural defences of the Cap Bon Peninsula by astute deployment of armour. All German and Italian troops in North Africa surrendered on the 13th.

Sheer numbers now dominated almost all phases of the air war. On 21 June Johannes Steinhoff led I./JG 77 off from Trapani to attack a force of B-17s. The interception was made late and Steinhoff alone succeeded in destroying one Fortress. Göring, noting that 100 Bf 109s had shot down one B-17 and seething with rage, fired off a signal on 25 June to the effect that one pilot from each of the fighter *Statteln* involved in the 21 June action would be court martialled for cowardice. It was that kind of reaction to dire operational problems that made many individual *Jagdfliegern* realise just what sort of totalitarian regime they were fighting for – provided of course that they were still ignorant of that fact.

A further blow to the Germans was the highly successful series of Allied landings in Sicily beginning on 10 July. In what was becoming a well-established practice, the landing beaches were subject to heavy naval bombardment before Allied bombers and fighters laid bomb carpets across the island's airfields. Bf 109s were hunted in the air and on the ground, craters and showers of debris left by tons of high explosive being as much of a hazard to their flimsy undercarriage as the actual detonations. Hundreds of German fighters were lost to this singular cause throughout the North African campaign, many aircraft that were not immediately destroyed having to be abandoned during the ebb and flow of the ground war. Air raids also shredded the nerves of the pilots, who were acutely aware that they could be killed or maimed without even getting airborne.

That the Allied aerial long arm was extending far into areas hitherto all but immune to attack was demonstrated on 1 August 1943 when a force of USAAF B-24s drawn both from groups that formed part of the Mediterranean-based 9th Air Force and the 8th in England struck at the giant Ploesti oilfields in Romania. In only the second raid on this target, the Liberators of the TIDAL WAVE force were harried by German, Romanian and Bulgarian fighters, the operation marking the combat debut for the Bf 109E in the hands of the *Vozdushni Voiski*, air arm of the latter nation. The Bulgarians were relatively ineffective against the B-24s, the bulk of their sorties being made by Avia B 534 biplanes and a handful of Bf 109Es which succeeded in shooting down two bombers. The American raid clearly highlighted a need for more improved interceptors. Immediately after the raid, Göring offered the Bulgarians 48 Bf 109Gs to supplement their Czech and French fighters.

Fighter *Gruppen* were moved from Sardinia to offer a token resistance to the Allied push across Sicily and pilots of JG 27, 51 and 77 tried their best to inflict casualties at minimum cost to their own units. A number of Allied machines were indeed brought down but it was prudent now for the survivors to seek the marginally safer airfields of southern Italy and a withdrawal across the Straits of Messina began on 14 August. By the time Messina fell to US troops on the 17th, the remaining fighter *Gruppen* in the Mediterranean had occupied Italian airfields and 60,000 German troops had narrowly avoided capture by reaching Italian soil.

With Allied bombing raids on the *Reich* demanding more and more experienced pilots for defence of the homeland, the *Luftwaffe* high command viewed the future defence of Italy as an increasingly lost cause and the remaining fighter *Gruppen* were progressively withdrawn; II./JG 27 left its Bf 109Gs behind to be used by JG 3, 53 and 77 which were based on the Foggia complex of airfields for the time being, while II./JG 51 also returned to Germany to re-equip. As events were to show, the Germans would not have to denude Italy entirely of effective fighters, for Bf 109s in Italian hands carried on the fight, ably filling a void where the *Luftwaffe* could not.

By carefully marshalling a handful of Bf 109s, the *Jagdwaffe* was able to offer a final challenge to the Allied landing in Italy on 1 September. II./JG 53 made short work of ten P-38 Lightnings on the 2nd, a combat that nevertheless saw the demise of *Oblt* Franz Schiess, who had 53 victories to his credit. The following day, Italy surrendered.

Chapter 6
New Threats

By the late autumn of 1940 the Bf 109F had entered service in small numbers initially with *JG* 51, examples of this early F-1 variant having also been issued to the *Ergänzungsgruppen* (reserve fighter training groups) of *JG* 26 by November. The former unit suffered what is believed to have been the first loss of a Bf 109F over the Channel when the machine piloted by *Oblt* George Claus, the 1. *Staffelkapitän*, failed to return after a combat with Spitfires on the 11th. Werner Mölders, who had not flown on the operation due to a severe bout of 'flu, was livid. Ignoring all protests, he took off for an immediate search for his close friend Claus, but to no avail. Claus had in fact survived to become a prisoner.

Losses sustained after the main daylight assault on England had stopped seemed that much more poignant and perhaps, wasteful, none more so than when very able *Kommandeuren* were the victims. On 28 November JG 2 flew an afternoon sweep over the Solent, led by the I *Gruppenkommandeur*, *Major* Helmut Wick. Surprising Spitfires of Nos 602 and 213 Squadrons, the Bf 109s scored two victories, one almost certainly falling to Wick for his 55th kill.

Other *Luftwaffe* fighter units clashed with the RAF on a day that saw considerable air action. II./*JG* 51 met No 501 Sqn and shot down one Hurricane while I./*JG* 26 fell on the climbing Hurricanes of No 249. *Oberst* Galland shot down P/O Wells before his unit engaged Spitfires of No 19 Sqn which chased the Messerschmitts and shot down *Fw* Kiminsky and *Uffz* Wolf.

On a second sortie, I./*JG* 51 made for Southampton, maintaining high altitude. Observing Spitfires, Wick and his *Stabschwarm* dived steeply. Wick carried out a text-book attack on one of the British machines, almost certainly that flown by P/O Baillon of No 609 Sqn, which fell in flames.

Oblt Erich Leie and *Lt* Fiby broke off and lost sight of Wick but the *Kommandeur*'s wingman, *Oblt* Rudi Pflanz, stayed close to his leader. Then a Spitfire attacked Wick's machine, causing what must have been fatal damage, as Wick promptly baled out. Seeing the parachute open, Pflanz wheeled around after the Spitfire, caught it and shot it down south of Bournemouth. It was later identified as almost certainly that of P/O John Dundas of No 609 Sqn, although his body was never recovered.

No 512 Squadron had meanwhile arrived to support No 609 but the Spitfires were themselves bounced by more Bf 109s. The Spits scattered and one fell. P/O Marrs pursued the Messerschmitt that had done this damage and in turn, shot it down, the pilot baling out before it exploded. In total the RAF lost eight Spitfires and four Hurricanes destroyed for three Bf 109s in half a day's combat in which there had been the opportunity for the relatively rare 'pursuit of the pursuer' to take place.

Helmut Wick's loss was mourned by JG 2, all the more so because he had had no need to fly that day. A strongly-worded grounding order had been despatched but did not apparently arrive at the Richthofen *Geschwader*'s base by the 28th. Such a loss was a final body-blow to the morale of the *Jagdflieger* which was smarting under the totally unjustified accusation of cowardice by their C-in-C. The force's relationship with Göring was never to be quite the same after the *Kanalkampf*.

Despite a modest production order (a total of only 257 aircraft according to one source) the Bf 109F-1 served with *JG* 1, 2, 3, 26, 51, 52, 53 and 54, but by no means were all *Gruppen* fully or even partially equipped as the sub-type was intended primarily for pilot familiarisation pending delivery of the similar Bf 109F-2. As there was a general slowdown in fighter operations during the winter of 1940/41, Bf 109F conversion training for the bulk of the *Jagdwaffe* could be conducted uninterrupted, although it took only a short time for the majority to master only slightly changed flying characteristics of the new model. Deliveries continued into the early part of 1941, most *Gruppen* not receiving their aircraft until the turn of the year.

In France, JG 1 had joined JG 2 and 26, both of which continued to fly the Bf 109E to maintain full operational strength while converting pilots to the F-1 on a rotational basis. JG 26 was in January 1941 preparing to detach 7. *Staffel* to operate from Sicily but an effective Western European fighter defence against incursion by the RAF had to be maintained as, with improved weather, the British would undoubtedly mount an offensive, although how substantial this would be, only time would tell.

The *Jagdwaffe* meanwhile began to fully re-equip with the Bf 109F-2 and F-4, the latter variant, which had followed the F-2 into production almost immediately, introducing the heavier but slower-firing (compared to the MG 151/15) Mauser 20-mm MG 151/20 centreline cannon. Good as this weapon was, tests of captured examples showed that it had few advantages over the standard RAF Hispano 20-mm cannon and that its elaborate electrical cocking system and disintegrating belt-type feed was probably more prone to malfunction. Nevertheless, the Germans manufactured the Mauser gun in quantity and it was used throughout the remainder of hostilities on both the Bf 109 and Fw 190.

It was in January 1941 that a number of Bf 109Fs suffered the aforementioned tailplane failure, the factory implementing a remedy in the form of four external straps. These were attached to fuselage frame 9, below the tailplane, although they did not appear on early production F-1s or F-2s. The modification was of course necessary on both sub-types and was probably retrofitted prior to internal strengthening being introduced during the course of F-2 production. Nevertheless, the tailplane stiffeners continued to appear on some early production F-4s.

Winter weather permitting, *Jagdwaffe* cross-Channel operations continued. Largely freed of the tiresome escort duty – although bombers and *Stukas* sortied against England in daylight when tactical conditions were favourable – the fighters flew *frei Jagd* sweeps and escorted Bf 109E fighter bombers. The latter duty, intended as only one-way cover, was more acceptable to the fighter pilots as the speed margin between a bomb-laden *Emil* and their own unencumbered machines, was minimal.

The *Jagdwaffe* did not have to wait long for a British cross-Channel offensive to materialise, as the RAF mounted Circus One on 10 January, the first of a long series of fighter/bomber sweeps. The German fighter force was the first line of defence against such intrusions although the most important targets were well defended by *Flak* units. From the British viewpoint, enticing the *Jagdwaffe* to fight served to even the odds in case the enemy attempted, once the weather improved, to renew the daylight bomber offensive on England. Nobody knew for certain how much of a possibility this was.

The few German pilots who rose to defend the first Circus target, the airfield at Forêt de Guines, were faced with a formidable number of enemy aircraft. Three Hurricane and three Spitfire squadrons escorted six Blenheims to the target and three additional Spitfire squadrons covered the exit of the main force over the Channel. The Blenheims duly bombed Guines and there was limited air combat with two Bf 109s being shot down. The German pilots claimed a Hurricane.

As the Bf 109E was giving way to the F, so the Hurricane was being supplanted in the first-line British fighter squadrons by the Spitfire. The *Jagdfliegern* were to intercept many more sorties flown by Hurricanes in the ground attack role, however. They found that the Bf 109F could hold its own even against the improved Spitfire Mk V although below-the-radar sweeps by small numbers of fighters could not always be intercepted. These latter were maintained in addition to larger scale sweeps and escorts to bombers – known as Ramrods – which had a specific target to attack. Otherwise, the RAF's 2 Group tactical bombers often acted as bait, hoping to entice the Germans up to do battle.

The British operations developed much as the Germans' own had done over southern England in 1940; the object now was a steady decimation of the *Jagdwaffe*, the disruption of communications and transport in north western France and damage to airfields and installations useful to the German war effort. A war of attrition therefore began, to ensure that the *Luftwaffe* was obliged to maintain substantial defensive forces on the Channel coast.

In their turn, assisted by an efficient radar and signals service, the *Jagdwaffe* positioned fighters up-sun to 'bounce' RAF raids which at first were always heavily biased towards fighters. The Germans could afford to choose their moment to attack, the high speed dive, fire and climb tactic remaining effective. If a dogfight developed, the *Jagdfliegern* enjoyed considerable advantage; pilots who were wounded or had to bale out were invariably safe from capture. Occasionally the combats took the fighters out over the Channel, although chasing enemy aircraft was not encouraged. The range of the Bf 109F, at 440 miles, was not greatly superior to that of the E model and was a primary factor in dictating the parameters under which the Germans fought.

While JG 2 and JG 26 maintained the main defence of what might be termed the European 'central front', German fighters were stationed on the peripheries in Holland and in Norway, where elements of JG 77 and *Jagdgruppe* Stavanger operated. The latter was one of a number of independent *Gruppen* formed during the war and which more or less remained operational in one area, with the advantage that pilots and ground controllers became thoroughly familiar with local conditions. They also came to know intimately the type of opposition they would be likely to encounter, the main flight paths adopted and the performance characteristics of enemy aircraft types.

The Bf 109s belonging to *Gruppen* based in northern climes were primarily sent up to intercept RAF anti-shipping aircraft and bombers flying 'cloud cover' sorties on days when they had an even chance of evading the Messerschmitts. While the types encountered were mainly mediums and two of the RAF's heavies, the Halifax and Stirling, JG 77 was also responsible for shooting down two of the RAF's Boeing Fortress Is before the experiment in using ultra-high altitude bombers in small, vulnerable formations, was terminated.

Among the German casualties suffered during this period was *Maj* Wilhelm Balthasar, *Kommodore* of JG 2. He was shot down and killed by Spitfires on 3 July 1941. More fortunate was *Maj* Rolk Pringel, *Kommandeur* of I./JG 26 who came down in southern England and was made prisoner.

The *Jabo Staffeln* of both JG 2 and 26 received the 109F-4/B, capable of carrying a similar range and weight of bombs as the E model, but enjoying a better performance. While *Jabo* attacks on mainland Britain were only occasionally effective in disrupting transportation and industrial output, they had an adverse effect on morale, as the bombing was often indiscriminate and could not always differentiate between military and civilian installations. On the other hand, shipping plying the Channel, Thames Estuary and coastal waters afforded the fighters far more worthwhile targets. JG 2's pilots concentrated on shipping targets while JG 26 worked along the coast.

Among the favoured targets were gas holders, those giant structures located on the outskirts of many large English towns and which were almost impossible to camouflage. Using the faster Bf 109F, the *Jabo* pilots adopted a flatter approach to their targets than before; the high speed dive from altitude was abandoned in favour of a low level run-in that helped to mask the attackers from radar detection for as long as possible.

Oblt Frank Liesendahl was one of the pilots who rose to prominence in 10./JG 2; his efforts added materially to British shipping losses which had reached about 63,000 tons by 26 July 1942. Liesendahl himself had been posted missing on 17 July and although single-seat *Jabo* losses rose steadily as the British defences improved, they remained an integral part of the *Luftwaffe's* daylight offensive against the British Isles and served to tie down a proportion of Fighter Command squadrons until the forward bases were lost after the 1944 invasion.

Having just taken up his appointment as *General der Flieger*, Adolf Galland was handed an exacting task timed to take place on 12 February 1942. This was Operation *Cerberus*, the daring movement of the *Kriegsmarine* capital ships

Scharnhorst, Gneisenau and *Prinz Eugen*, from their blockaded German ports to Norway's deep water fjords – after traversing the Channel. A huge fighter umbrella was required and by deploying almost his entire first-line fighter strength in the West and making up the numbers with aircraft from *Jadgschüle* 5 near Paris, Galland assembled a force of 252 Bf 109s and Fw 190s.

Some of these saw combat with RAF and FAA machines when it was belatedly realised what the Germans were up to. Bad weather all but prevented many enemy aircraft from finding the ships and the operation was an outstanding success, the *Jagdflieger* losing only a single Bf 109 and three Fw 190s in return for destroying 42 of the enemy.

Spring bought more action against the RAF, which continued its cross-Channel offensive. Spitfires were now the main fighter type opposing the *Jagdflieger* and in June 1942, the first Mk IX was issued to No 64 Sqn at Hornchurch. Developed mainly to counter the Fw 190, the Spitfire IX was the equal of the Bf 109F and G and while the *Jagdgeschwadern* in France suffered no alarming rise in casualties as a result, the days when their pilots could expect to meet inexperienced enemy pilots flying inferior aircraft, were virtually over.

Six months or so after the 'Channel Dash' the *Jagdwaffe* in France faced a new threat. On 17 August the first US 8th Army Air Force heavy bomber raid was mounted against the marshalling yards at Rouen-Sotteville. The modest force of B-17E Fortresses from the 97th Bomb Group was not attacked by the German fighters and only on the 8th's third foray to the continent was one Fortress damaged, in an Fw 190 attack. Four further missions went unchallenged by the *Jagdwaffe*, which bred a false sense of security in the minds of some American commanders, unsure of how successful their bold venture might prove to be.

Two days after the initial appearance of American four-engined bombers, the *Jagdwaffe* was heavily engaged in countering the Allied invasion 'dress rehearsal' at Dieppe. Another example of the Allies testing the well-prepared German defences too early with too small a force,

Operation JUBILEE was a disaster. It included a very strong fighter element with more than 750 Spitfires, Typhoons and Hurricanes participating. Around 220 Bf 109s and Fw 190s rose to challenge this mighty force which in addition to single-engined fighters, involved Beaufighters and medium bombers.

Many of the *Jagdfliegern* had never seen so many 'peacocks' eyes' in the sky at once and although the enemy squadrons operated over the beach-head in relays, the air seemed full for the duration of the ill-starred amphibious landing. Up against odds of more than three to one, the German fighters nevertheless shot down 100 Allied fighters.

The lion's share of victories went to *JG* 2 despite four of the RAF squadrons operating over the invasion force flying the Spitfire IX. It is doubtful that participating *Jagdfliegern* had time to notice or recognise the subtle difference between this and the less capable Mk V.

High-flying *Gustav*

The German casualties over Dieppe had included a Bf 109G-1 of *JG* 26. Operated by 11 *Staffel*, a special high altitude unit formed within I *Gruppe* primarily to develop tactics against high-flying US heavy bombers, this machine was probably that flown by *Oblt* Johannes Schmidt, who had 12 victories. Powered by the Daimler-Benz DB 605 engine, the G-1 featured cockpit pressurization and an associated canopy sealing system with some additional framing plus some airframe structural strengthening but was, in its earliest variants, outwardly similar to the Bf 109F.

Initial examples of the Bf 109G-1 reached first-line units in June 1942, production ending the following month after 167 aircraft had been completed. *JG* 2 had in fact been the initial recipient of the G-1, this special *Staffel* being joined by 11./*JG* 26. The G-1 was also issued to 3./*JG* 1, *JG* 51 and *JG* 53, albeit in small numbers as the Bf 109G-2, concurrently built by the Erla and Regensburg plants, was the first of the main production variants of the *Gustav*.

Fw 190s finally opened the *Luftwaffe*'s air combat victory tally against the 8th Air Force on 6

September, one machine from the 97th Group going down. To keep the Germans guessing as to the primary target for the heavies, the 8th sent the 15th Bomb Squadron's A-20s to make a few holes in Abbeville, home of II./JG 26. The B-17s meanwhile made a two-pronged attack on St Omer and Méaulte.

In England, the build up of the strategic force committed to a policy of precision bombing of enemy targets in daylight went on. Four groups of B-17s and one of B-24 Liberators arrived during September, these to make good transfers of units from the 8th to the 12th Air Force in the Mediterranean theatre. Inclement autumn weather prevented a number of planned missions going ahead although the 8th gave a clear signal of its intentions on 9 October when 108 B-17s and B-24s struck the Fives-Lille steelworks in Belgium. This 'maximum effort' was not to be equalled for some months and while the *Luftwaffe* reaction was strong, only four bombers were lost. Again it was Fw 190s that did the damage, one pilot opening the *Jagdwaffe*'s account with the B-24. These early heavy bomber incursions were invariably challenged by the Fw 190s of JG 2 and JG 26.

While the Americans stacked up their bomber formations so that the high squadrons maintained around 25,000 feet, both German fighters were, not surprisingly, capable of operating well above this. The Bf 109G-1 possessed an absolute ceiling of 39,700 feet compared to 37,403 ft for the Fw 190A, which led to a marginal preference for the Bf 109 if ultra-high altitude interception was necessary. This was rarely the case as the bombers' most favoured operational altitudes generally ranged between 21,000 to 27,000 ft. Although the B-17F could operate up to 37,500 ft, the B-24D's maximum was considerably lower, at 28,000 ft. On combat missions, Liberator formations usually adopted height bands between 19,000 and 23,000 ft.

The good altitude performance of the German fighters was advantageous in that they could usually position themselves well above the bomber stream and carry out high speed attacking passes as soon as the B-17s and B-24s entered their sector. Radar tracking and the monitoring of numerous radio test transmissions from the bomber stations in England before a mission, gave the Germans ample time to prepare a hot reception. On certain days, the unmistakeable white condensation trails created in the cold, thin air by a big formation was an additional aid to the defences.

Throughout the remainder of 1942 the German fighter force found more than enough action against RAF medium bombers and fighters as well as the American heavies, but there was little immediate change in operational tactics. There was no denying however that the Americans' precision attacks, aimed at a single airfield, shipyard or factory complex was cause for concern, for such methods seemed to offer a more direct threat to German war industry than the haphazard night bombing raids then being mounted by the British. The US bombers usually maintained close formation to achieve a good concentration of their bomb loads on the target and they carried an exceptionally heavy armament of around a dozen machine guns apiece. Clearly, the *Luftwaffe* and the *Flak* arm had to make such a daylight bomber offensive so costly that the Americans abandoned it ...

By early 1943 the *Jagdwaffe* was poised to deliver an extremely punishing series of blows to the 8th Air Force heavy bomber offensive, which was slowly but surely growing in scope and strength. Many of the most important targets for the 8th Air Force lay deep in Germany and initially, the incursions were limited to coastal areas. Even to hit these from bases in East Anglia the B-17 crews faced a long flight across the North Sea and occupied Europe. They could thus be subject to attack for hours of flight-time over hostile territory.

For their part, German fighter pilots were now faced with destroying the most heavily-armed bomber force the world had yet seen, even though it was by that time, only modest in size. While the number of *Flak* guns that defended the most important targets were steadily increased, the most effective method of bringing down the heavies, was a well armed fighter. This was particularly true of damaged machines that had cleared the target area but could still be destroyed well away from a *Flak* zone if fighters were up. For the *Jagdwaffe*, North-western Europe was fast becoming an air front as potentially dangerous as an invading ground army.

JG 2 and 26 were consequently reinforced by pulling units out of other war theatres for temporary periods. No new fighter *Jagdgeschwadern* with the 'traditional' full strength of three component *Gruppen* were formed at this time, the *Luftwaffe* planners preferring to maintain existing formations at full strength or to add a fourth *Gruppe*.

As the importance of countering the US daylight raids increased, the additional *Gruppen* began forming specialised anti-bomber *Staffeln*. In the *Jagdwaffe* these units mainly flew single-seat fighters, but twin-engined types and fighter versions of bombers, particularly the Ju 88, were specially adapted to this highly demanding task.

Shooting down *Viermots* (four-engined bombers) was significantly more difficult than any other aircraft type the *Jagdflieger* had yet faced but the ruling that a pilot must have destroyed 20 enemy aircraft in aerial combat before being eligible for the coveted *Ritterkreuz*, remained unchanged for the time being. It came as a shock to pilots making their combat debut in the West that this seemingly modest figure was far harder to attain than it had ever been elsewhere. Later, the scoring system for decorations was modified.

The deadly work inevitably brought to the fore pilots who were more able to bring down the heavies than their colleagues. These 'four-engined specialists', such as Walter Osau, Georg-Peter Eder and Heinz Bar, sprinkled throughout the fighter *Gruppen*, were individuals who possessed that extra degree of fine judgement, superb eyesight and lightning-fast reflexes – not to mention raw courage – that enabled them to hack down the well-armed heavies from close range. Equally importantly, they inspired their fellow pilots to emulate their success.

The Germans could usually hit the Fortress formations at least twice, sometimes more if the raid was a 'deep penetration' one. Bombers would invariably be lost both en route and out of the target area, despite their use of different courses, attacks on more than one target and diversionary formations. Very few missions from England managed to return entirely unscathed, the weather occasionally creating the exception to the rule when the fighter force failed to make contact.

For many months, accurate bombing required visual sighting and 8th AAF losses were occasionally out of proportion to the 'poor to minor' damage inflicted on the target. The Germans became adept at repairs so that some industrial plants that appeared to have been heavily bombed were able to maintain output, albeit at a reduced level. Both the B-17 and B-24 had limited bomb capacity both in terms of total weight and size and many targets required repeat visits to keep output disrupted. Total destruction was a goal that was rarely achieved.

Despite the increasing effort being put into the American daylight bomber offensive and the demonstrably greater destruction being wrought by the RAF at night, the *Luftwaffe* was exacting a steady toll of heavy bombers. On 27 January 1943, the 'Boeings' had bombed Wilhelmshaven, their first target in the Reich.

The interception of this and other American raids was often undertaken by units equipped with the Fw 190A series, the overall performance of which was superior to that of the Bf 109G, which really required additional weaponry to be effective against heavy bombers. Any performance penalty imposed on the lightweight airframe slowed the Bf 109 and/or made it marginally less manoeuvrable. In comparison the 'Butcher Bird' had been designed around integral cannon with substantial hitting power, making it the obvious choice for the bomber interception role. At altitude however, the lighter 109 had few equals.

Output of both fighters rose during 1943 despite disruption caused by the bombing but Fw 190 production was never planned to equal that of the Bf 109. It was therefore decided to largely direct the bulk of Fw 190 deliveries to units stationed in Western Europe, with the Bf 109 serving as the primary fighter type elsewhere. Germany's waning military position meant that this ideal could not always be followed and ultimately, the Bf 109 more or less retained its 'front line' status on all war fronts.

Having been weaned on the Bf 109, some pilots did not take to the Fw 190 – indeed a handful positively threatened to desert if their beloved *'Beule'* was taken away! The handling characteristics of the two fighters was markedly different and it was sometimes sound practice to retain the familiar but older type rather than have a *Staffel* or *Gruppe* effectively out of action for pilot conversion training. Also, the early Fw 190s had their share of teething troubles and technical problems, which forced operational plans to be changed, pending modifications. In that situation, a *Jagdgeschwader* would have one or more of its component *Gruppen* flying both types, a fairly common occurrence that was to continue for the duration of the war. A number of special sub-variants of both fighter types were developed and these were usually available to give individual *Jagdgeschwadern* considerable freedom of action to adapt to changing operational requirements.

Some two months after B-17s first penetrated *Reich* airspace, the AAF was able to escort the heavies as far as the range of its new single-engined fighter would allow. This, the P-47 Thunderbolt, was to further wear down the *Jagdfliegern* in 1943 for although the big American radial did not have the range to escort its charges much beyond Paris, many a German pilot met his end when challenging it.

The American heavies had in fact been partially escorted since their earliest days in England but it soon became clear that short-legged RAF Spitfires were not the answer. Being able to shepherd the heavies little further than the Dutch coast caused no end of frustration in British as well as American circles, but nothing practical was done to redress the situation. All the British fighters could do was to operate from the most forward UK airfields and cover the bombers as far as Brest or Cherbourg and that was pushing their range capability to the limit. The worst aspect of this problem was that there was little fuel reserve for combat.

Much hope was then placed in the P-38 Lightning with its twin-engined safety factor and heavy, concentrated armament. But technical problems so plagued this sleek twin that AAF chiefs decided as an alternative, to try the only other American type extant with anywhere near the performance and firepower to survive in the skies over Europe.

Consequently the P-47 became the first of the true escort fighters; its heavy armament and incredible diving speed made it a deadly antagonist to the lighter German fighters – but the heavies were increasingly obliged to fly beyond its protection to hit targets deep in Germany.

The P-38 of course later claimed the distinction of becoming the first US fighter to fly over Berlin, on 3 March 1944 but a year earlier, few people could predict with confidence that its fundamental problems could be solved quickly. The 8th Air Force wanted long range fighters in substantial numbers immediately.

By basing interceptors along the bombers' most favoured routes, the *Jagdwaffe* offered as much of a threat as its own fighter numbers would allow. And the losses suffered by the American and British bomber commands rose steadily as both the day and night fighter defences of the *Reich* were expanded. By day the *Jagdwaffe* fought hard to stave off the US bombers with less than 250 fighters.

German aircraft losses were still being replaced easily but a creeping paralysis for the *Luftwaffe* was the difficulty in filling the shoes of those pilots lost in combat with men of equal skill. Training was only keeping pace with demand and no substantial reserve was ever built up. This problem had not grown really acute in any one theatre by 1943 but the writing was on the wall – Allied strength would build to the point where whatever force was needed to defeat Germany would eventually be made available.

By mid-year Germany could not militarily extend her influence or expand her industrial base; her raw material supply was finite and the gradually spiralling level of Allied interference would eventually create shortages that could not be overcome. In the aviation industry, research was begun into using other materials, particularly wood, as an alternative to aircraft grade aluminium. For the Bf 109, a wooden vertical tail section did prove to be a viable alternative and a full wooden mainplane was built but never introduced in production.

Night moves

In addition to the burgeoning American daylight bomber offensive, the *Luftwaffe* was, by mid 1943, having to contend with the heavy night raids by RAF Bomber Command. The devastating attack on Hamburg on the night of 25/26 July, during which 'Window' was used for the first time to blot out the vital radar guidance of conventional night fighters, led to the first deployment of single-seat fighters in the *Wilde Sau* role. These target defence night fighting sorties did not rely on radar and they met with some initial success, particularly by the Bf 109Gs flown by *JG* 300. This unit was established at the behest of Hajo Herrmann to boost the efforts of the conventional *Nachtjagd* formations. Although it fell some way short of being the ideal night fighter, due mainly to its limited size being unable to accommodate much extra equipment, the Bf 109G-6 was nevertheless adapted to take the FuG 360Z *Naxos Z* passive homing device. Thus equipped, it carried the designation Bf 109G-6/N and was delivered to I./*JG* 300.

Along with similarly-equipped Fw 190s, the Bf 109G-6/N also underwent operational trials in the hands of pilots of *NJGr* 10 – but what was really wanted at that time was an effective answer to the de Havilland Mosquito, the scourge of the night skies over Germany. All but immune from interception on a regular basis due to its speed and manoeuvrability, the Mosquito problem haunted Hajo Herrmann and others dedicated to finding a counter to it.

The FuG 217 *Neptun-J*, the radar that evolved from the FuG 216 tail-warning set, appeared to offer a solution and by early 1944 a total of 35 Bf 109s and Fw 190s had been fitted with it for testing, again mainly by *NJGr* 10. As *Yagi* aerials could only be positioned on the Bf 109's fuselage, the more convenient wing leading edges being occupied by slats which prevented a set of 'stag's antlers' aerials from being carried, the interceptions made by converted day fighters fell largely to the Fw 190A which could carry the extra equipment.

These sorties however remained more or less experimental, neither the Bf 109G-6/N nor the single G-4 used to test FuG 217 apparently becoming operational with *NJGr* 10. *NJG* 11 did operate unmodified G-6s, G-10s and G-14s as pursuit and area defence night-fighters for a period that lasted until January 1945 whereupon these same machines were fitted with bomb racks for nocturnal ground-attack sorties, a role which took on an increasing importance at that stage of the war.

Limited success attended the *Wilde Sau* sorties flown in subsequent months but the high command knew the value of a second line of defence to the twin-engined Ju 88 and Bf 110 night fighters, and the small force established through Herrmann's enthusiasm, remained in being. A degree of rivalry sprang up between the *Nachtjagd* and the *Wilde Sau* pilots of *JG* 300 and this denigrated into outright acrimony; both sides regarded the other with some disdain, the single-seater pilots being considered irresponsible daredevils by the regular night-fighter crews who had achieved their success by more patient and methodical means.

The initial *Wilde Sau* formation, I./*JG* 300, suffered a high rate of attrition owing to a hasty period of training foisted on its pilots and ground crews but this at least was intended to be a self-sufficient formation with its own aircraft and personnel. II./*JG* 300 at Rheine had the added difficulty of being in effect a stepchild of III./*JG* 11, which was tasked to support a parasite formation and provide it with aircraft. The day fighter units had trouble enough without these newcomers regularly wrecking their Bf 109s after their dangerous nightly forays. III./*JG* 300 at Oldenburg fared little better. Shadowing II./*JG* 1, its pilots met with much the same reaction.

Although the *Wilde Sau* concept was far from perfect, the failure of the *Nachtjagd* (through no fault of its own) to forestall the night bombing led to its expansion. Hermann Göring decreed the establishment of *JG* 301 and *JG* 302 in September 1943, both units initially receiving enough aircraft and pilots to form one *Gruppe* each. The former unit was based in Bavaria while *JG* 302 began operations from Brandenburg. Hajo Herrmann was given a new command, 3 *Jagdivision*, to direct his force.

Swiss adventure

While there is no disputing the *Luftwaffe* C-in-C's influential part in the Bf 109 story, his direct orders only rarely concerned Germany's principal fighter. In April 1944 however, an incident in Switzerland gave the *Reichsmarschall* little choice but to exercise his authority. On the 28th a Bf 110G night fighter made an emergency landing at Dubendorf after pursuing RAF bombers. Officials in Berlin almost had apoplexy, for the aircraft not only carried the then-secret SN-2 radar but was fitted with the deadly obliquely-mounted '*Schrage Musik*' (jazz music) guns which could be so devastating to RAF aircraft. There was no way the Germans were going to allow the Swiss to retain the aircraft and everything was done to retrieve it, including, if all else failed, the carrying out of an armed raid led by the crack commando, Otto Skorzeny.

The Swiss stalled, duly examined and photographed the Bf 110's secrets and generally awaited developments. The incident reached the highest diplomatic levels and culminated in a deal whereby Germany would sell the Swiss 12 Bf 109G-6s for 500,000 gold francs each if the Bf 110 was destroyed. After some anxious moments, when a *Luftwaffe* expert confirmed that the radar had been removed and reinstalled, the aircraft was duly blown up on 18 May. The Bf 109s arrived on the 19th and three days later, Göring received a cheque for six million Swiss francs.

The *Reichsmarschall* was never to hear the end of this story, but he might well have been aware of what actually took place. Finding that they had been partially duped in that all the Bf 109Gs had been fitted with old, worn-out engines, the Swiss demanded compensation. And six years after the war ended, Messerschmitt and Daimler-Benz were obliged to pay up. On the positive side, this strange deal gave the Swiss *Flugwaffe*'s fighter force a useful and economical life extension, enabling Messerschmitts (or 'Daimlers' as they were nicknamed) to remain part of the air force inventory well into the postwar period.

Chapter 7
Russian Roulette

Hitler's greatest military crusade began in the early hours of 22 June 1941 when massive Luftwaffe air raids heralded the last major German surprise attack on another country in WWII. Three *Wehrmacht* army groups enjoyed the support of three *Luftflotten* to carve out North, Central and Southern spearheads, each with its own objectives under the plan for Operation *Barbarossa* which committed some 3,000,000 men to a 'decisive' campaign designed to totally subjugate Russia and turn it into a mere adjunct of the *Reich*.

In support of the respective land armies, the *Luftwaffe* fighter order of battle included the Bf 109Es and Fs of I., II. and III./JG 54, plus the unit's *Einstatsgruppe* and II./JG 53 under *Luftflotte* 1 to protect the bombers assigned to the northern force, while *Luftflotte* 2 had the Bf 109E strength of II and III./JG 27, the Bf 109Fs of all four *Gruppen* of JG 51, II./JG 52 and I and III./JG 53, plus two *Gruppen* of ZG 26 and SKG 210 with Bf 110s. *Luftflotte* 4 directed I., II. and III./JG 3; I./JG 52; II. and III./JG 77; E/JG 77 plus I (*Jabo*) LG 2, all of which flew the Bf 109F except JG 77 which retained the E model, many of which had been passed on by JG 54 shortly before the campaign in the East began. In addition, *Luftflotte* 4 could call upon the strength of four Rumanian fighter units and twelve bomber squadrons.

Three days after the Russian attack opened,

Similarly marked to the machines in the previous photograph is a Bf 109E-3 of III./JG 52. Numerous *Emils* remained in use for the early stages of *Barbarossa*. (*Holmes*)

Meanwhile, the *Emil* also served well and long in the far north where *JG* 5 spent much of the war. *Uffz* Arthur Beth of 8. *Staffel*, still flying an E-4 in summer 1943, then had 16 victories. (*J V Crow*)

Finland came into the war on Germany's side, provoked both by Russian bombing and bitterness over the terms of the 1939-40 Winter War. Five Finnish fighter units went into action, mainly to provide air support to the Finnish army's advance down the Krelian Isthmus to recover territory ceded to the Russians. Refraining from joining the Germans in major joint operations against the Russians, the Finns nevertheless represented a useful buffer force in the far North for which relatively few German troops were needed.

Irrespective of which model of Bf 109 the *Jagdwaffe* flew on the new Eastern Front, the neutralisation of the Red Air Force was achieved at an unbelievable speed; at every point, the Russians' ability to react to the invasion was at best weak and at worse, too late. *Voennof Vosdushyne Sily* (*VVS*) aviation units had either received no advance warning of the sudden

Fighter casualties on the Russian front included this F-2, 'Black 1' which appears to have been brought down by striking aerial wreckage, judging by the spinner damage. Russian troops are examining the remains.

A Soviet guard on 'Yellow 5', another otherwise anonymous Bf 109F shot down early in *Barbarossa*.

shattering of the Moscow-Berlin non-aggression pact, or scant orders as to what to do in the event of being attacked. Surprise was achieved almost everywhere and the *VVS* paid the price of having a greater part of its front line strength deployed in Western Russia, well within range of *Luftwaffe* medium bombers and dive bombers.

Few of the *Jagdflieger* had seen anything like it. Their Bf 109s swept aside all opposition and Red Air Force aircraft simply fell out of the sky under the onslaught – those that is, that had been able to take off. Many enemy aircraft had been disabled on their airfields by hundreds of SD-2 anti-personnel bombs dropped in the early raids. German troops consequently found vast numbers of Soviet aircraft that appeared relatively undamaged but had been disabled by shrapnel from the deadly bombs. While SD-2 bombs were more commonly conveyed to their targets by Ju 88s and He 111s, *JG* 77 was among the fighter units that had Bf 109E-4/Bs adapted to carry them.

It was over Russia that German fighter pilots, old heads and novices alike, brought the deadly art of air combat to awesome new heights. Just as the Red Army could not check the sheer speed of the ground advance, so *VVS* tactics were often found wanting in the face of the *Jagdwaffe*'s excellent air discipline. The Russians could however almost afford to lose aircraft even at such a vast rate, as they had some 7,000 machines of all types available for service when *Barbarossa* began.

The scale of air fighting breathed new life into the race to see which German pilot in this war would beat von Richthofen's 80 kills in World War I. When the campaign in the East began, Adolf Galland of *JG* 26 was in the lead with 69 kills, followed by Werner Mölders of *JG* 51 with 68. The way now seemed clear for Mölders, for Galland remained in France.

Mölders' four victories on the first day of *Barbarossa* brought an immediate award of the Swords to his *Ritterkreuz mit Eschlaub*. Other pilots soon equalled his multiple score in a single day, the weather being fine and highly conducive to the task in hand. Russian losses spiralled to reach 322 aircraft shot down by fighters and *Flak* units plus 1,489 destroyed on the ground. Very quickly the *Jagdflieger* achieved air superiority over a distance of about 250 miles into Russia.

Such a large scale operation could not be undertaken without losses and while the number of German fighters shot down was comparatively small, they inevitably included highly experienced leaders, some of whom had opened their air combat record in Spain. Men such as Wolfgang Schellman, *Kommodore* of JG 27, who had 26 victories and *Hptm* Heinz Bretnutz of II./JG 53, who had 27, would be sorely missed. Both were shot down on 22 June. Schellman crashed after attacking an I-16, the enemy fighter breaking up and showering his Bf 109 with fragments that damaged it fatally. Forced to bale out near Grodno, Schellmann was probably executed by the *NKVD*, a grim portent of what German flyers could expect at the hands of the Russians, even this early in the conflict. Right from the start, German aircrew were treated harshly by the Russians, with little distinction being made between these men and the ground troops. Bretnutz was wounded and had a leg amputated but died on the 27th.

Not all captured airmen were executed but those who survived invariably faced a long and arduous incarceration. *Oblt* Kurt Sochatzy, *Staffelkapitan* of 7./JG 3, was shot down by the Il-2 he attacked on 3 August and had the far from unique experience of being imprisoned until 1949. Unbeknown to the German flyers, the Russians considered all military personnel on the opposite side to be 'war criminals'. That fact alone mitigated against many pilots who fell into Russian hands and appeared to give front-line units the power of summary execution without any form of 'trial'.

Accidents also continued to take their toll of fighter pilots. *Hptm* Lother Keller of II./JG 3 lost his life on 26 June and was replaced by *Hptm* Gordon Gollob. Much more would be heard of Gollob, an ex-*Zerstörerflieger*, while fame for exploits of a different nature had followed the new *Kommandeur* of I./JG 53 when he took up his post in July. *Hptm* Franz von Werra, the only German pilot to escape – albeit indirectly – from British captivity, had finally returned to take the controls of a Bf 109. Shot down in a Bf 109E in 1940 while serving with JG 2 on the Channel coast, von Werra had jumped from a train in Canada, crossed to the then-neutral USA and finally returned to Germany. Destined not to survive for long, he was killed in October 1941 when his Bf 109F-4/Z went into the sea off the Dutch coast after I *Gruppe*'s transfer from the East that summer.

'Assi' Hahn, *Kommodore* of JG 2 perched on a Bf 109E-4 fitted with additional pilot's head armour. Of note is the lack of a white background to the Richthofen *Geschwader*'s script 'S' badge. (*Crow*)

Some *Jagdflieger* who were veterans of Spain became re-acquainted with the I-16 *Rata*, the type ruefully remembered as the one Russian fighter that could, with a good pilot in the cockpit, give their Bf 109s considerable trouble. Such veterans needed few reminders of the almost incredible manoeuvrability and heavy firepower of the I-16, which remained in service with the *VVS* in some

RUSSIAN ROULETTE

numbers. Not that outstanding Russian pilots had as yet had much of a chance to shine – but these were early days.

The German fighter units that occupied forward airfields found many wrecked and disabled *Ratas* and I-153s, plus examples of the newer Russian fighters, the LaGG-3 and Yak-1. Combat on the Eastern Front had brought the total of enemy fighter types to have seen action against the Bf 109 since the war began to well over a dozen, of Polish, French, Dutch, British, American, Yugoslavian and Russian origin. It was some measure of Willy Messerschmitt's design ability that his brainchild had managed – in some cases only by a slim margin – to match, if not outfly and outfight them all.

Combat in Russia also found the Germans facing an enemy air force that was in many ways similar to their own by being almost 100 per cent tactical in nature. The *VVS* was an integral part of the Russian military forces and had not, in common with the Germans, developed a force of long

As Russian resistance stiffened the *Luftwaffe* had also to abandon aircraft on the Eastern Front; these hangared Bf 109Fs were probably part of II./JG 51. (*via R L Ward*)

An F-2 with a *Panzerglas* external bullet-proof windscreen and belly mountings for a bomb rack under Allied evaluation.

range bombers, although the four-engined Pe-8 was in service in small numbers.

As the campaign developed, the *Luftwaffe* was unable to maintain total air supremacy all along the front – it was simply too vast an area for the number of aircraft the Germans had available – but the encircling of entire Russian armies meant that huge tracts of territory were quickly made 'safe'. The *Jagdwaffe* moved forward and occupied dozens of Russian air bases, the ubiquitous Ju 52s supplying their immediate requirements in fuel, spares and replacement personnel.

By late June 1941 the race to beat von Richthofen's WWI record had been won by Werner Mölders. And on 15 July, as if to emphasise that in the current conflict German fighter pilots could almost regard a score of 80 victories as a routine achievement, Mölders scored his 100th victory. This feat brought 'Vati' the Diamonds and promotion to *Inspektur der Jagdflieger*. While this latter honour meant at very least a curtailment of operational flying, Mölders continued to fly sorties until he could no longer delay taking up his staff appointment.

Hptm Walter Oesau, *Kommandore* of JG 3, got his 80th kill the same day as Mölders' 100th and was posted from the Russian front to take command of JG 2 in France. Eight more fighter pilot recipients of the *Ritterkreuz* were announced during July 1941, a reflection not only of the meteoric scoring rate of German pilots in the East but the consequentially heavy attrition brought about by the vast number of operational sorties being flown from dawn to dusk: by mid-July 1,284 German aircraft had been shot down or damaged, which meant that almost all the original strength of the *Luftwaffe* for *Barbarossa* had required replacement in less than four weeks.

The speed with which the *Jagdfliegern* were building up their personal scores had never previously been possible, a fact that led the authorities to revise the requirements for the award of the *Ritterkreuz*. Presented only for outstanding acts of bravery in battle, Germany's highest decoration would inevitably be devalued in status if too many men were eligible to wear it – if fighter pilots continued to accrue victories at the present rate, the wearing of the *RK* would become almost *de rigeur* in all the fighter units!

The answer was to make the medal harder to win. Accordingly, the number of victories necessary for a pilot flying in the East to be eligible was raised from 20, then 25 and to 40 by the end of 1941. In all cases these were minimum scores and the award was by no means automatic, even when these figures were substantially exceeded.

If the Bf 109 had begun *Barbarossa* with an outstanding reputation, the opening months of the war in the East more than enhanced it through the brilliant achievements of its pilots. Their exploits made excellent material for the German press and radio broadcasts and the wartime *Luftwaffe* magazine *Signal*, which splashed the exploits of the leading *Experten* across numerous issues. Fanmail

The rudder of the Bf 109F denoting the 112 victories scored by *Hptm* Heinz Bar of *Stab* IV/JG 51 pictured at Kertsch in Russia on 27 June 1942. (*Crow*)

from home made life at the front more bearable although at that stage of the war, few German fighter pilots had cause to complain. With the campaign going so well, most of them wanted to continue to accrue victories and to speed the final defeat of the Soviets.

One of many pilots who made their name on the Russian Front was Walter Nowotny. Having joined 9./JG 54 in February 1941, the young *Leutnant* opened his score on 19 July by shooting down three I-153s near the island of Osel. Forced to ditch his Bf 109 in the Baltic, Nowotny took to his dinghy and spent three days paddling with his hands before reaching dry land. Returning to his unit, Nowotny built upon his initial success over the next year and had shot down 54 enemy aircraft by 4 August 1942.

As one of the new arrivals in the East in July, Lt Hermann Graf was, at 30 years of age, somewhat senior to his contemporaries in 9./JG 52. Having spent some time on training duties, Graf had had no previous chance to score victories but from that point on, few *Jagdflieger* were able to match his scoring rate. Graf's fellow pilots in III./JG 52 were more than pleased to be once again in the thick of combat and flew Bf 109F-4s from Bielaia-Zerkow during these early months.

While his victory tally was still modest, Graf met Werner Mölders during one of the latter's frequent visits to the front. The Crimea was about to erupt into a German offensive aimed at Kersch and Sevastopol and Mölders wanted full information on respective German and Russian air strength and while talking to Graf he predicted a great future for him. As Mölders' judgement of pilots was almost legendary, Graf was naturally pleased.

Mölders' prediction was given substance on 23 October when *Fw* Ede Duhn, the 23-victory *Staffelkapitän* of 9./JG 52, was killed in combat. Hermann Graf was appointed *Staffelkapitän*. It appeared that being denied the services of experienced pilots was a cross that III./JG 52 was to continue to bear, for barely a month later, Gunther Rall's Bf 109F-4 was shot up by a Soviet fighter. Attempting to crash-land, Rall's machine made a very hard landing, causing him to fracture his spine, an injury which hospitalised him for nine months. Graf went from strength to strength and was awarded the *RK* on 24 January 1942, having shot down 42 enemy aircraft in the space of five months.

JG 3, 51 and 52 were then heavily engaged in operations around Rostov, primarily to protect German troops attempting to cross the River Dneiper. The Russians flung in their bombers to stop this dangerous development and they were swatted like flies by the ever-efficient *Jagdfliegern*. JG 51 alone claimed 77 aircraft shot down in two days and while the component *Gruppen* of JG 3 had notched up the *Geschwader*'s 1,000th kill on 30 August, JG 51 had by then achieved a war total of 2,000 victories. JG 54 also did well and among its accolades was the award of the *RK* to Hans Philipp of II *Gruppe* on 24 August, by which time his score had risen to 62.

Such milestones were not reached without losses and the casualty list included *Hptm* Hermann-Freidrich Joppien of JG 51, killed in action on 25 August and *Oblt* Kurt Sochatzky of JG 3. The latter pilot had had the not uncommon experience of having been shot down twice before – the difference was that on both previous occasions he fell behind Russian lines. The third time, 3 August 1941, Sochatzky was made prisoner when his Bf 109 crashed near Kiev. He would not be the last *Jagdflieger* to go down when attacking the notorious Il-2 and the manner of his loss was also far from unique. Pressing in close to his victim to put telling shots into the heavily-armoured aircraft, Sochatzky saw the *Schturmovik* break up but could not avoid the tumbling wreckage, a piece of which tore off one of the Bf 109's wings.

After numerous combat reports had been scrutinised, the *Jagdflieger* came to realise that the only way the Il-2 could be destroyed was by attack from below, where the aircraft had little armour protection. If shells penetrated its vulnerable coolant radiator, the Il-2 was all but doomed. Zooming up from below also avoided getting too close and exposing the fighter to fire from the rear gun; this was not the only successful attack angle, but boring in very close from behind or above put the German pilot at considerable risk.

Enemy fire, either from aircraft or the ground could of course, easily wreck the Bf 109's own cooling system. Such was always a risk with liquid-cooled engines and one that both sides had to contend with. Fortunately for the Germans, the majority of Russian combat aircraft were powered by in-line engines and a pilot with a good eye and a fair degree of marksmanship could knock them down with a few well-aimed bursts from cannon and machine guns.

As September 1941 brought the first tangible signs of winter, the German campaign still pressed forward towards its triple objectives. The *Jagdwaffe* managed to maintain local air superiority on almost all sectors although Russian strength in the air was noticeably increasing. A great many *VVS* fighters challenged the air cover (mostly provided by JG 77) for Army Group South's drive into the Crimea, where the Russians secured local air superiority for the first time.

Newly promoted to command II./JG 3, *Hptm* Gordon Gollob received the *RK* on 18 September, the date of his 42nd kill. All *Gruppen* in Gollob's *Geschwader*, plus JG 51, then became embroiled in Army Group Centre's Operation *Taifun*, the encirclement of a Russian army group in the Smolensk area. On 19 October Army Group South plunged into the Crimea and met stiff resistance which took ten days to break.

October 1941 also saw Mölders' place at the head of JG 51 filled by *Maj* Friedrich Beckh. This opened up the race for top scorer once again and on the 24th, *Oberst* Gunther Lutzow became the second *Experte* in the *Luftwaffe* to reach 100 kills. Multiple victories in one day were not unusual, but nine was indeed something to celebrate as JG 3 duly did when Gollob totalled up his score at debriefing on 18 October. By the 26th this highly respected officer's score had risen to 85.

All fighter units in the East continued to fly the Bf 109F and G and even the E model had not been entirely relegated to second-line duties, for shortly after the start of the Russian campaign the *Luftwaffe* had been joined by a number of Axis air units, from Italy, Hungary, Croatia and Slovakia. The bulk of these was concentrated on the southern front under *Luftflotte* 4, which suddenly had a motley collection of biplanes as well as more modern types, to integrate into operations. Those units equipped with Bf 109s initially flew the *Emil* and this variant tended to be retained for some time after *Luftwaffe* formations had received the F and G.

Bf 109Es were also flown by pilots of Spain's *Escadra Azil*, the unit including a number of pilots who were civil-war veterans. Attached to III./JG 27 and otherwise identified as 15. *Staffel*, the Spaniards were led by *Cdt* Angel Salas Larrazabel, who had scored 17 victories over his homeland and was to add a further seven while serving in the East. Both Slovakian and Croatian fighter pilots operated as small national units in *Staffel* strength under the jurisdiction of JG 52, while the Spanish unit subsequently transferred to JG 51, similarly retaining its 15. *Staffel* identification.

When the Russian winter of 1941/42, the severity of which had not been equalled for many years, stopped the German spearheads within sight of Moscow, the *Jagdflieger* enjoyed a welcome lull in operations. The bitter weather on the Eastern Front brought some changes to the composition of the fighter force, mainly to provide reinforcements for other war theatres. It was the Mediterranean that received the bulk of these, including the three Bf 109-equipped *Gruppen* of JG 53 and II. and III./JG 27.

When operations had to be flown, the *Luftwaffe* groundcrews found themselves having to run engines in temperatures few of them had ever experienced. Day after day the thermometer dropped even further below the zero mark – and yet the Russians continued to fly, seemingly untroubled by the weather. It was true enough that the *VVS* was more than able to cope with such conditions which had over the years, become commonplace for the time of year. Shortly before the *Wehrmacht* attempted a last abortive push to take Moscow on 25 November, the thermometer registered 15 degrees of frost . . .

Interrogation of Russian aircrew prisoners helped to solve any problems the Germans experienced in starting recalcitrant engines. The answer was to thin the engine lubricating oil with petrol; oil congeals in sub-zero temperatures and needs reliquifying. This startling practice, with its seemingly high fire risk, worked well enough, for when the engine started, any remaining fuel quickly evaporated. An alternative was to ignite a tray of petrol under the Bf 109's engine compartment to thoroughly warm it prior to attempting a start. A different solution was found with freezing gun mechanisms. Simply by washing off all the lubricating oil and protective grease (which also froze in low temperatures) either with petrol or boiling water, enabled the guns to function perfectly.

Stalemate

With the 1942 spring thaw, the fighting in Russia renewed and was marked by success and failure for both the *Wehrmacht* and the defenders; none of the primary German objectives had been secured although Leningrad in the North was isolated and under siege and the Baltic was a virtual German lake. With better weather, plans were made to capture the Caucasus oilfields, eliminate Leningrad and complete the task of wresting Sevastapol from the Russians. All operations required substantial *Luftwaffe* support with local air superiority secured by the *Jagdwaffe* so that the *Kampfgeschwadern* and *Stuka Gruppen* could attack targets largely unmolested by Soviet fighters.

All along the front the Bf 109 continued to be the scourge of the *VVS*; JG 51 being among the outstanding units in terms of victories. And JG 52 which had hitherto enjoyed relatively little success, now began to make its mark in the East. Flying the Bf 109F, all three *Gruppen* were based in southern Russia to support Crimean operations. JG 77 also did well as the fighting in the Crimea grew in intensity. But it was becoming clear that despite appallingly high losses, the *VVS* had recovered; every aircraft the Germans shot down in combat was seemingly replaced by two or three as production far beyond the range of *Luftwaffe* bombers began feeding new and improved machines to the front line units.

In May 1942 JG 5 was formed from cadres provided by the Petsamo *Jagdgruppe* and the Kirkenes *Zerstörergruppe*, to guard the northernmost areas of *Reich* territory. Its Bf 109s covered Norway and the Baltic region and was destined, with some transfers, to remain in that area for much of the war.

On 14 May *Oberstlt* Adolf Dickfeld of III./JG 52 shot down nine Russian aircraft to bring his personal score to 90. Milestones in the *Jagdwaffe*'s combat record in Russia were constantly reached and surpassed; 'Macky' Steinhoff's 100th victory came on 31 August. Each individual pilot's victory accumulated on the *Abschlüss* (operational record) of his parent unit and on 4 November 1942 Hans-Joachim Heyer's 53rd victory brought the total for JG 54 to 3,000. That victory was often tempered by losses was shown when Heyer was killed five days later when his Bf 109 collided with a Russian fighter.

Late 1942 saw the debut of the Lavochkin La-5, LaGG-3, and the first examples of the Yak-7/9 series. The Russians were unique amongst the Allies in fielding nine first-line fighter types, the largest number of any of the combatants. Without exception, all of them were simple but rugged machines well able to stand up to the rigours of operations under all weather conditions.

Fortunately for the *Jagdflieger*, the Russians' doctrine of almost totally deploying airpower as a tactical weapon meant that the *VVS* generally operated at low to medium (below 10,000 feet) altitudes, flying as an integral part of the ground units to which they were attached. Had the Germans attempted to fight the Russians at the latter's favoured heights, fighter pilot casualties would undoubtedly have been higher. But by adhering to their proven dictate that to have height enables greater control of the battle area, the *Luftwaffe*'s fighters remained a formidable force, even when they were outnumbered or were themselves attacked during bomber and *Stuka* escort sorties in direct support of the *Wehrmacht*.

Time and time again the *Jagdwaffe*'s *frei Jagd* patrols would come upon Russian aircraft, sometimes in huge gaggles, sometimes in ones or twos. The enemy pilots frequently failed to maintain a good look-out, and would invariably lose one or two of their number, picked off by the fast diving Bf 109s. There often appeared to be no set pattern in the way the Russian aerial patrols operated, for they too flew free hunting sorties designed to whittle down the *Luftwaffe* whenever its aircraft appeared. German pilots quickly came to appreciate the vastness of the Eastern Front and identifying the few man-made features, particularly railway lines, could make the difference between a successful flight home and the horror of following a reciprocal course which could easily take an aircraft into the endless 'sea' of featureless steppe until its fuel ran out.

In October 1942 Erich Hartmann joined 9./JG 52 then based at Soldatskaya on the river Terek. Greeted by the *Gruppenkommandeur*, the newcomer fervently hoped he could make his mark. Just how well he was to do so, none could imagine at the time but Hartmann would operate for three weeks before his first victory was confirmed, a not unusual occurrence for many pilots who went on to greater achievements.

By the autumn of 1942 the *Wehrmacht* began to be faced with the increasingly difficult task of interpreting conflicting orders from Berlin. Hitler's indecision over the priority goals of the campaign in the East brought repeated modification of the original tactical plan for securing Moscow, Leningrad and the oil fields of the Caucasus. This hesitancy gave the Soviets a slim breathing space and made a wholesale collapse far less likely.

With overstretched supply lines and a finite number of replacement tanks and men to make good the already substantial casualties, the *Wehrmacht* was obliged to switch large forces from one sector to another and the more or less straight front line that marked the limit of its advance to the Volga began to bulge dangerously under Red Army pressure. A similar juggling of resources faced the *Luftwaffe*.

Rapidly changing operational areas became very familiar to the *Jagdwaffe* during this period and JG 77 was a case in point. Having fought alongside JG 3 and been based in the Kiev area until the end of 1941, I *Gruppe* transferred to the Mediterranean in July 1942. II and III *Gruppen* remained on the southern sector of the Eastern Front until the latter, plus the *Gruppenstab*, was moved north to Leningrad in September 1942.

Within weeks, JG 77 in its entirety moved to the Mediterranean, the experience of this *Jagdgeschwader* not being too dissimilar to that of JG 53 which was also to spend the mid-war years fighting in North Africa. A nucleus of pilots from 1./JG 77 were transferred to form I *Gruppe* of JG 4, a new *Jagdgeschwader* which was initially based in Rumania flying Bf 109s to protect the Ploesti oil fields.

Combatting Russian fighters wherever they appeared, the *Jagdwaffe* steadily depleted the VVS while flying escort sorties to *Stukas*, bombers and transports and trying to avoid enemy AA fire which was invariably accurate and a considerable hazard over some sectors of the multiple fronts. *Gruppen* were moved throughout the *Luftflotten* areas of responsibility to assist the ground forces by strafing enemy troops, road transport and trains in an attempt to consolidate German positions before the winter struck again. In front of Stalingrad, the city which by December, had become virtually the sole focus of Hitler's Eastern campaign, the *Wehrmacht* could not break the Russian defence.

Keeping fighters inside what had become a giant pocket with the enemy pressing in on all sides, the *Jagdwaffe*'s strength dwindled as hundreds of sorties were flown to protect the transports tasked with supplying the trapped Sixth Army. In mostly appalling weather, the fighters could not prevent the unfolding disaster and when Gumrak, the last remaining airfield in German hands inside the Stalingrad ring had to be abandoned, the end was in sight.

That was on 22 January 1943; the last fighter to evacuate the pocket was a Bf 109G-2 of JG 3, the sole survivor of a local airfield protection *Staffel* originally based at Pitomnik. Comprising six pilots, all volunteers, this small formation fought hard to offer German troops some protection from Russian ground attack aircraft and subsequently moved to Gumrak in the closing weeks of the battle for the city. II *Gruppe*'s Kurt Ebener did well, claiming 33 kills in the Stalingrad area before the final curtain which came down on 2 February when *Feldmarschall* Paulus surrendered the shattered remnants of the Sixth Army.

After Stalingrad, substantial German gains in the East became increasingly rare. Having broken the myth of German invincibility and received an enormous moral boost, the Red Army garnered its strength for the forthcoming task of winning back all lost territory. The major German ground defeat had been accompanied by a substantial loss in aircraft, particularly in precious transports. And although the *Jagdwaffe* had made it as

One of the most effective *Jagdgeschwadern* on the Russian Front, JG 54 forged a fine reputation for aerial prowess. A Bf 109G of 8./JG 54 prepares to taxi out in wintry but clear weather. *(Barbas)*

difficult as possible for the Soviet airmen to share in the glory, this could not disguise the fact that Stalingrad had also drained the fighter force of about 100 aircraft.

By early 1943 the Bf 109G-6 had begun to reach front line units in Russia generally to supplant the older model *Gustavs*; early production G-6s had been delivered to units based in the Mediterranean and Western Europe, where the opposition was considered to be that much more dangerous. Apart from the numbers of combat aircraft the Soviets could deploy, the *Jagdgruppen* in the East remained a small but formidable force and could afford to await the arrival of a new Bf 109 sub-type in quantity in the spring.

Still considered to be lightly armed for combat in Western Europe, the G-6 offset this disadvantage to some extent by the substitution of the cowling-mounted MG 17 machine guns with the heavier MG 131. In order to accommodate the larger spent cartridge belt chutes for these weapons, the G-6's rear cowling had to have a distinctive bulged fairing on each side. The '*Beule*' (bump) was born.

The Bf 109 also remained the primary German first-line fighter type in the East although the number of Focke-Wulf 190s steadily increased after some initial technical problems had been rectified. As regards fighter armament, the Germans were not too disadvantaged in combat with the Russians, despite the generally perceived light weight of fire available to the Bf 109G. The Fw 190 was invariably superior in this respect – although the principal Soviet fighters had even lighter armament than the Bf 109F in order not to compromise manoeuvrability and speed.

At their favoured low level operating altitudes

the Yak and LaGG series relied on fuselage-mounted cannon and machine guns. To keep wing loading down, few Soviet fighters, in common with the Bf 109, mounted integral wing guns. All Russian fighters also exhibited an outstanding rate of roll and generally excellent manoeuvrability and it was this latter quality that the *Jagdflieger* needed in their fighters, gunpower being not quite so critical.

There now began a series of fighting retreats by the *Wehrmacht*, with the German fighter force covering the abandoning of the Caucasus region by February 1943. Kharkov changed hands twice and although a number of key cities were lost, the German supply lines were considerably shortened while those of the Red Army were lengthening. The mid-year tank engagement at Kursk, designed to straighten the front line and eliminate a dangerous bulge caused by Soviet pressure, resulted in a further massive defeat for the Germans. Land battles on such a scale inevitably drew in substantial air support and as at Stalingrad, the huge tank battle claimed peripheral fighter casualties when the *Jagdfliegern* flew hundreds of successful sorties against the omni-present *VVS* ground attack aircraft.

Hitler had staked so much on the *Panzers* being able to inflict massive losses on the Soviets at Kursk that the failure of the battle was a decisive turn of the tide. Russian casualties had been substantial but their ability to replenish losses now completely eclipsed that of the Germans. A retreat of major proportions was now the only hope of saving manpower and equipment.

When the German pull-back began in the face of the spring 1943 Soviet offensive, the *Jagdwaffe* found itself reoccupying some of the same airstrips it had used during the 1941 advance. Few of these boasted concrete runways or dispersals or more than the barest facilities although fighter units could operate with very few home comforts provided the transports could fly in fuel and supplies.

The Bf 109 coped remarkably well with the primitive conditions, despite accidents continuing to take their toll of its weak landing gear. Occasionally, fine weather could present more difficult landing conditions than mud or snow for dust and the presence of small stones could cause an uncontrollable slide during the landing run. During the winter months the mainwheel covers were often removed to prevent a build-up of packed snow from freezing the wheels solid. Under such conditions the groundcrews performed their usual miracles in getting cold engines started, utilising the methods previously described if the unit did not enjoy the luxury of portable heaters. Patching battle damage and refuelling and re-arming became as second nature to the hard working 'blackmen', the groundcrews without whose dedication the *Luftwaffe* could not have operated.

By September 1943, German tanks were back on the Dneiper; Smolensk in the south had been retaken by the Russians, as had Kharkov and their tanks and infantry doggedly pressed forward all across the front. Hitler's insistence on a series of fighting withdrawals instead of a general pull-back to establish a solid defence line in order that the armies could regroup, adversely affected the *Luftwaffe*, which had to carry out its exhausting fire brigade sorties against numerous danger points.

The *VVS*, true to its *raison d'être* as an aerial army, stormed over *Wehrmacht* positions with impunity, the casualties inflicted by German fighters now appearing to make very little difference to the number of pilots and aircraft available to the enemy. That did not stop the *Jagdflieger* shooting them down in steadily increasing numbers.

Chapter 8
Three Front Disaster

By late 1943, the German position on all major war fronts was growing increasingly precarious; the *Jagdwaffe* had in a remarkably short time been transformed from an offensive to a defensive force, with the high command demanding ever more pilots to defend the homeland against incursions by Allied medium and heavy bombers. And while ground attack and reconnaissance were roles of continuing importance, both of them utilising the Bf 109 as well as other types, it was bomber interception that increasingly absorbed fighter units based in the West – in fact this duty affected all front line pilots, for a posting to the *Reich* could seemingly be received at any time. Allied pressure made the continued defence of the Mediterranean all but a lost cause.

For the most part the front line units could expect their lost fighters to be replaced by almost exact duplicates of those they were already flying – no new conventional types had been developed early enough to enable a lengthy conflict to be fought with other than improved versions of the 'first generation' of German fighter – the Bf 109. Many new designs were under test and in the advanced design stage, but the war would have to last some years before they could realistically be expected to equip front line *Luftwaffe* units.

Only the advent of the Fw 190 had really

Although believed faked, this print of *Wilde Sau* Bf 109Fs with 'yellow 2' of an unknown II *Gruppe* in the foreground gives the feel of single-seat fighter operations against RAF night bombers. (*Holmes*)

improved the fighter situation. Having entered service some seven years after Messerschmitt's single seater first flew, Kurt Tank's excellent design was that much more modern and it had, unlike the 109, considerable development potential. The all-important heavy armament now required to deal with well-constructed and heavily defended American bombers could be accommodated that much more easily by the Fw 190, for although heavy guns and ammunition had the inevitably detrimental effect on performance, these would have been ruinous on the Bf 109.

Messerschmitt, preoccupied with numerous new projects including the radical Me 262, was never able to initiate a major update of the Bf 109 airframe. It is even doubtful whether such would have made a significant difference to the basic design, as fundamental revisions could not have been achieved without some disruption of the production lines at a time when output of fighters was becoming vital.

That is not to say that the 109 was not improved as the war progressed but the advantage of a more powerful engine was increasingly offset by heavier airframe and armament weight. Changes under war conditions could only be limited in scope and did not extend for example, to a new wing, strengthened to take integral cannon and with the wing slats deleted to improve low speed manoeuvrability. The modifications that were introduced addressed some of the basic design's long standing drawbacks (not in themselves critical), including the degree of vision from the heavily framed cockpit canopy.

The cockpit of all the early 109s was criticised (particularly in Allied evaluation reports, which did not really count) for a lack of a clear view, although the records of countless pilots proves that they quickly overcame this hindrance. Indeed, it was said that a framed canopy appeared to offer the occupant some degree of psychological security. Despite the fact that a 'frameless' canopy gave an unobstructed view of the world, there was, until a man grew used to it, an irrational fear of an adversary seeing in that much more easily. Irrational or not, this fundamental need for security, even to hide, stemmed from basic human behaviour and was far from unique to Messerschmitt pilots!

The Erla works developed a new cockpit hood which, while hinging to the starboard side in the same manner as before, drastically reduced the heavy side and top framing which had been in vogue since the Bf 109E-4. The *Erla Haube* (hood) was generally fitted as standard on all late production Bf 109Gs and Ks and could also be retrofitted to earlier models with little difficulty, the overall dimensions of both framed and clear canopies being similar.

While light airframe weight had always been one of its advantages, the Bf 109 soon faced the all-too common compromise situation of most WWII single-engined fighters in that it had ideally to maintain a good performance with firepower heavy enough to destroy opposing aircraft. Better armament tended to mean extra weight, quite apart from any other additional equipment. And a fighter the size of the Bf 109 had to do this with the utmost economy in ammunition expenditure, bearing in mind the limited space available for cannon and/or machine gun rounds.

In the Bf 109G, the weight spiral that threatened to nullify other attributes the design might have possessed, was already manifesting itself. In fully loaded configuration, the G-6 tipped the scales at 6,940 lb, which was 547 pounds more than that for the F-4. This translated, for example into a climb rate that was slower and a service ceiling below that of the earlier model. As mentioned previously, the only way that Messerschmitt could endow the *Gustav* with the heavy firepower required to bring down four-engined bombers and armoured ground attack aircraft such as the Il-2, was to add extra guns as *Rüstzustand*.

In addition, the Bf 109G series introduced the capability to carry and fire a pair of Wgr 21 air-to-air rockets launched from metal tubes attached to the wing underside. A makeshift weapon, this was based on the army's 21-cm *Nebelwerfer* (smoke dispenser) 42 rocket mortar and consisted of a 248-lb rocket with a 90-lb warhead. Theoretically quite effective for bringing down bombers, these weapons had no inherent guidance, although the warhead could be preset to detonate, via a time fuse, at 600 to 1,200 yards from the launching point. The fighter pilot had to

Bf 109G-1s after factory roll out, all bearing the *Stammkennzeichen* which were usually removed prior to delivery to a front line unit.

allow for an appreciable loss of trajectory when lining up his target and due to their being 'single round' weapons, rockets were not considered a great advantage over guns apart from the fact that they could be launched well outside the range of defensive guns on heavy bombers, thereby reducing the risk to the interceptor.

In line with well-established German practice, this policy of adding weaponry freed the airframe proper from incorporating weighty equipment that might only be needed for certain specialised roles. It was a philosophy that generally worked well throughout the industry but in the case of the Bf 109, which had originally benefitted greatly from Willy Messerschmitt's lightweight design, relatively little extra weight was allowed for. In the Bf 109's case, adding additional underwing weaponry was a compromise in comparison with the strides made in international fighter design under the stimulus of war.

Front line pilots however, tended to be fairly traditional in their attitude to their aircraft and often downright suspicious of anything 'new fangled' – they knew full well that the Bf 109G was the equal of most comparable single-seat fighters, that it was reliable and could be flown to known limits; it was a useful weapons platform and it would often bring its pilot home even though it may not have emerged from combat completely unscathed. Comparing those known attributes to some new type promising much but largely untried in the combat arena, many pilots would have opted to remain with the Bf 109 – which is exactly what many of them did for as long as was possible.

In order to destroy the maximum number of heavy bombers at minimum cost, the *Jagdwaffe* adopted a range of intercept techniques, all of which had merit and were adaptable enough to allow telling attacks to be made on individual combat boxes or flights. There was little need to stress the technique necessary to destroy a straggler, i.e. a single damaged B-17 or B-24, as the fighter pilot invariably had the advantage which he could exploit. A cripple was however not always the easy victim it may have appeared; few German pilots would minimise the risk of facing the veritable battery of heavy machine guns mounted in every Fortress or Liberator, irrespective of its condition. Those who were careless could still be shot down by a few well-aimed bursts.

The exact number of German fighters destroyed by bomber gunners will probably never be known with any degree of accuracy, although individual American crewmen were decorated for their prowess behind the sights of a single or twin 'fifty-caliber'. A number of men certainly achieved the status of ace by destroying five German aircraft, for the heavy Colt-Browning M2 could be a lethal weapon in skilled hands. And if there were 500 of them blazing away at a few German fighters, the 'lacework of fire' as one *Jagdflieger* described it, was bound to hit home sooner or later . . .

The actual confirmation of kills made by American bomber crews tended to be clouded by the massive over-claiming which, had it been

'Black 1', a Bf 109G-2 of II./*JG* 54 pictured in Finland during a detachment.

accurate, would have destroyed the entire *Luftwaffe* fighter force many times in the space of very few bomber missions. The unimaginable confusion of a big air battle led to many spurious kill claims being made in complete good faith by bomber gunners. As mentioned before, a contributory factor, one revealed early in the war, was the distinctive and highly visible engine exhaust trail left by both the Bf 109 and Fw 190 during their power dives. To many a harassed American gunner such a sight indicated a sure victory ...

Occupying huge areas of sky, both horizontally and laterally, the separate USAAF bomber boxes demanded a series of approaches by the German fighters. Depending on the weapons the fighter carried and the position of the opposing forces, attacks were often made by small numbers – even single German fighters waded in on occasion, despite the ever-stressed need that each leader should be covered by his 'wooden eye', his wingman.

Ruse attacks also paid dividends. Examples included a *Schwarm* of Bf 109s or Fw 190s approaching a bomber box from the rear. Starting their target run at 1,200 to 1,500 yards out, the four fighters approached in line-abreast formation, selecting one bomber as the target. At about 800 yards the outside pair broke away to port and starboard while the remaining two bored in. One more break at 700 or 600 yards would see one more fighter break away, leaving the remaining one to make the actual attack.

The object of this manoeuvre, ideally carried out in conjunction with a second *Schwarm*, was to confuse the bomber gunners, who could not be certain which fighter was about to open fire. Gunners who tracked the fighters which broke away momentarily exposed their aircraft to greater hazard from the ones that continued on their original flight path. And at typical closure speeds, all the *Jagdflieger* needed was a split second gap in the web of fire.

The 'scissors' and the 'roller coaster' were also effective in dividing the gunners' attention. The former had a pair of fighters diving in trail from above while others approached frontally on the same level as the combat box. On the breakaway, each diving fighter would climb or dive while the head-on element would vary its exit after attacking.

Fast and effective was the roller coaster. It involved a fighter initiating a dive from 4,000 to 2,500 yards out, the pilot aiming to pass under the formation. Just when the bombers thought they were safe from that particular quarter, the fighter pulled up, fired and climbed away, maintaining his airspeed. Ideally used when the bomber crews were further distracted by decoy fighters orbiting to port or starboard just out of gunnery range, the roller coaster demanded split-second sighting and firing if the *Jagdflieger* was to achieve a kill, particularly as it had to be carried out on the lead bomber box with clear sky in front to give the fighter room to climb away or to stall out and dive.

To assist the *Jagdfliegern*, detailed scale models of the American bombers were constructed with the maximum fields of fire from each power turret or hand-held position indicated by lengths of wire. These were designed to be studied by front line pilots to improve their attack technique but the American bombers were well protected and heavy fire could be expected from all angles. While the *Jagdfliegern* had a theoretical advantage in being able to vary the speed and direction of

One of many distinctly-marked Bf 109s flown by JG 54 on the Eastern Front, this F-2 was part of 7. *Staffel*. (Bernd Barbas)

their approach – they had to get in close to be sure of a kill. The element of surprise hardly existed.

The majority of *Jagdflieger* fully realised the responsibility vested in them in reducing the agony of their fellow Germans who were being bombed. A secondary factor was that scoring kills would bring not only the admiration of their fellows but essential experience, enabling the whole exacting business to get marginally – and only marginally – easier. If that led eventually to the honour and pride in wearing the coveted *Ritterkreuz*, everyone was aware of what it had taken.

The *Jagdgeschwader* holding the line against the American bomber offensive claimed some outstanding successes during 1943 which, had they occurred earlier in the war, might have induced a temporary lull or brought about a drastic rethink in the Allied camp. But by October, the combatants were committed to total war and even the combined loss of 120 plus B-17s from the two 8th AAF raids on Schweinfurt and one on the BFW plant at Regensburg, while damaging to German and American morale (for different reasons), caused only a small slackening of the pace. Within days the bombers were back over Germany, the previous casualties and the weather having in the meantime brought about some reduction in the number of aircraft the *Jagdfliegern* had to confront.

Throughout the conflict, Allied bomber leaders had to estimate losses in terms of a percentage from the total force despatched. As the number of effective sorties rose, so the percentage invariably adjusted and they rarely exceeded the 'unacceptable' ten percent even though the loss of hundreds of experienced crewmen at a time was grievous and appeared to some individuals to be biased towards some groups more than others.

But the Schweinfurt and Regensburg raids were something of a watershed; the American champions of daylight precision bombing had finally had to concede that the spiralling casualties inflicted by the German fighter force could no longer be tolerated: the bombers had to have 'all the way' fighter escort. Unfortunately for the *Jagdwaffe*, by the turn of the year the P-51B Mustang was all but ready to provide the kind of long range bomber escort that no other air force had tried or even dreamed to be possible.

A type not entirely unknown to some *Luftwaffe* pilots, the Mustang had been tested at Rechlin in July 1942 when the first example of the Allison-engined version had been captured. This RAF Mk I probably gave the Germans little inkling of the potential of the basic design when fitted with a new engine – but on 11 January 1944, the P-51B powered by the Rolls-Royce Merlin, made its debut over Europe.

Improved Gustav

During the summer of 1943, *Luftwaffe* fighter units in the West began receiving the Bf 109G-14. This model had started life as an attempt to bring together all the modifications that had been made to the G-6 and thereby introduce a 'standard' Bf 109 which would certainly have assisted Messerschmitt's network of sub-contractors, saddled as they were with an ever-increasing list of modifications. The attempt largely failed. The G-14 and the G-10 (which was built out of numerical sequence) were both produced in as many sub-variants as before, a fact that does not actually appear to have slowed the production rate at any of the plants!

Some standardisation was achieved with the Bf 109G-14, however. The *Erla Haube* cockpit canopy replaced the older-style hood and the *Peilrufanlage* radio navigation aid, which utilised a direction-finding loop mounted on the aircraft's dorsal spine aft of the canopy, became a useful recognition feature. Externally the early G-14 was otherwise similar to late-production G-6s which had been retrofitted with both these features.

Early production Bf 109G-14s were powered by the DB 605A engine which retained the forward fuselage contours of the G-6. But similarly to the G-6AS, the G-14 more commonly had the DB 605 AS engine which incorporated a larger supercharger. This in turn altered the fuselage nose contours and in effect smoothed out the distinctive bulges over the gun breeches that had hallmarked the G-6. In addition to the revised forward fuselage panels, the G-14/AS was

THREE FRONT DISASTER

fitted with the larger fin and rudder, an enlarged supercharger air intake and a VDM airscrew with broader blades.

With its improved high altitude performance, the Bf 109G-14/AS was issued to the majority of the *Reich* defence *Gruppen* while they were flying, for the most part, the G-6. Both sub-types were operated until the end of the war.

Also built out of sequence was the two-seat Bf 109, the G-12. Its surprisingly late appearance in the spring of 1944 was another result of Germany's 'short war' policy which failed to anticipate anywhere near the number of fighter pilots that would eventually be needed. Standard trainers were plentiful and it was believed that while the characteristics of the Bf 109 could obviously not be duplicated exactly, there was little need for a trainer based on the fighter. This was largely true all the while the instructors were highly experienced men with hundreds of combat hours in their log books. But as these individuals began to be lost in action, the young trainees were faced with the dilemma – how could they be expected to achieve much by being transferred straight from flying a basic trainer to the cockpit of a Bf 109 or Fw 190 in a front line unit?

The answer, in part, was to provide a few hours' dual instruction on the Bf 109 or Fw 190 and both fighters were produced in two-seat versions. The Bf 109G-12 was the generic designation for the two-seat series, conversion being made from G-2s, G-3s, G-4s and G-6s. About 500 conversions were originally planned.

Installation of a second fully-instrumented cockpit necessitated reducing the size of the fuselage fuel tank from 400 to 240 lts. The fighter version's 'L' shaped tank gave way to a flat tank lying along the bottom of the fuselage. Range was consequently restricted and the endurance was only 35 minutes. For longer training flights, the G-12s normally carried the standard 300-ltr belly drop tank. The rear cockpit of the G-12 had clear triangular panels built out from the side to give the instructor (who normally occupied the rear seat) some forward view over the pupil's shoulders. Each canopy section opened separately, there being a fixed centre-section between the two.

That improved Bf 109s could not in themselves solve the *Luftwaffe*'s predicament was demonstrated almost immediately the P-51B appeared on escort missions to 8th AAF bombers. A catalogue of appalling losses started with the January 1944 figure of 233 Bf 109s and Fw 190s removed from first line strength. February was even worse at 355 with an additional 155 damaged. The *Jagdflieger* were still hitting the bombers but now the task was that much more difficult and carried greater risk, as the USAAF fighter escort had invariably to be out-foxed or out-fought over their own airfields in France.

In addition to the 8th AAF from England, the Germans had had, from 1 November 1943, to contend with a second American strategic bombing force, the 15th based in the Mediterranean. This latter's operations could not simply be challenged by neatly allocating fighter *Gruppen* to the South, as the necessary back and forth shuttling of units and experienced leaders had perforce to continue. This only resulted in denuding one front to the detriment of another and increased the risk of *Experten* and novice alike succumbing to increasingly long odds: the more an individual flew, the more likely he was to be lost either in action or to the not insubstantial risk of accident, one perhaps compounded by the sheer fatigue of operations. With relatively few periods of leave, some

As the war progressed, pilots of the RAF Enemy Aircraft Flight were able to log flight time on captured Bf 109s, including this F-2, formerly of a *Jabo* unit. (*Merle Olmsted*)

German fighter pilots spent literally years in the front line, constantly exposed to danger. And yet it was the *Experten* who held the force together, acting as cement to bind the rest who strived to make the grade and emulate those who made the majority of kills.

During the fifth spring of the war, Germany was about to be sandwiched between two Allied armies whose military air strength had far outstripped her own. Although the Eastern Front lacked a heavy bomber dimension and in general terms the *Jagdwaffe* fared considerably better against the Russian tactical forces than their colleagues operating in the West, combat attrition was an ever-present factor. In the East, the *frei Jagd*, long since having had to be abandoned on any scale in Western Europe, still resulted in a substantial score for many pilots – the down side was that they could easily fall victim to the more intensive conditions within a few days or weeks if they were transferred to the West.

Europe's winter weather had been a blessing for the Germans before and that of 1943/44 was no exception. Both sides welcomed the slower pace of operations in the face of freezing fog, sleet and snow, all of which reduced visibility to levels far too dangerous for flying, even in a war. Mission postponements became commonplace but the lull at least gave the 8th Air Force the chance to check out new fighter groups which arrived in England. By the time the weather improved, the *Jagdwaffe* could count on three groups, the 354th, 357th and 4th opposing them in Mustangs.

That German intelligence was never very far off the mark when it came to identifying new arrivals was evidenced by the fact that the 4th Group's base at Debden, Essex, was raided on 19 February. This was barely a month after the group converted to Mustangs and could hardly have been coincidental. Little damage was done by the bomber crews, who missed Debden.

In March 1944 Berlin became the focal point for a series of 8th AAF raids with the escort seeking to draw the *Jagdfliegern* up. *Oberstl* Egon Meyer, *Kommodore* of JG 2, had on 5 January become the first pilot to achieve a score of 100 victories in the West. Meyer's total included 25 heavy bombers, which more than merited the award of the Swords. This was duly announced on 2 March – the day Meyer died in combat with P-47s of the 56th Fighter Group.

Traumatic news the following day was that American fighters had flown over Berlin for the first time. It was doubtful if many citizens witnessed the P-38s overhead, as Lt Col Jack Jenkins' 55th Group was way above the cloud cover, watching out for the *Luftwaffe*. Jenkins himself had his aircraft's troublesome Allison engines to contend with, not to mention some fifteen German fighters which chased but failed to catch him.

On 6 March, all hell broke loose over Berlin as the *Jagdwaffe* tried every tactic in the book to shoot down the American bombers. The number of participating Bf 109s and Fw 190s reached 400 and the slaughter on both sides was appalling. The defenders brought down 69 B-17s – but at a cost of 81 German fighters.

By then a new breed of German pilot was taking over from the many old heads who had fallen in aerial combat – but some of their adversaries were also new to combat, for the 6 March Berlin mission was the first big one for most members of the 357th Group. Each side's achievement was remarkable under the circumstances, particularly as the Germans were hampered by a dogged adherence to traditional air combat tactics. Manoeuvres such as the half-roll and dive, the 'split-S' – which often proved fatal if the pursuing enemy fighter pilot was quick off the mark, were still widely favoured. Both the P-51 and P-47 could easily overhaul the Bf 109 or Fw 190 in a full throttle dive, but some German pilots seemed never to have absorbed this lesson. The Thunderbolt was particularly lethal in this respect, as few other Allied or German aircraft could match it.

While victories over American fighters were gratifying, the *Jagdfliegers*' sole task was to shoot down heavy bombers if they were to stand any chance of at least reducing the weight of bombs that would surely fall on the selected targets. But the odds against them penetrating through to the bomber boxes were lengthening, particularly when it was realised that the AAF was capable of putting up more escort fighters on a single

THREE FRONT DISASTER

mission than the entire combined strength of the *Jagdwaffe* in the West.

After a heavy defeat at the hands of fighters and bomber gunners, the latter proving lethal on the numerous occasions when Bf 109s or Fw 190s stayed in range just a microsecond too long, the Germans had little choice but to impose a temporary stand-down in order to marshal enough aircraft for subsequent operations in any strength and to provide replacement pilots with a few hours' indoctrination into the current situation, all of which took precious time. Such an enforced lull occurred on 22 March when on one of the final Berlin missions of this period, American bomber crews returned home to report seeing hardly any German fighters in the sky.

Berlin raids were nevertheless suspended after March, the AAF considering the cost of 141 B-17s and B-24s, plus 56 fighters, to have been disproportionately high. The German capital represented a long range mission and it was invariably well defended. In their turn, the German units defending the capital had lost 168 fighters. The air defence of Germany benefitted from a worsening ground situation on other fronts, as fighter units which could do nothing further to prevent the inevitable collapse of ground forces, were pulled back to bolster home defence.

The *Reich* became an established war theatre in its own right and early in 1944 the *Reichvertidigung* (Reich Defence) identification marking had begun to appear on the fighters in the form of coloured fuselage bands for each participating unit. And home defence saw some units which had traditionally fought in other theatres back on German soil, among them *JG* 27. In April 1944 the *Geschwader* was based on Austrian airfields, its Bf 109Gs emblazoned with

Victim of the fighting over the Kuban in May 1943, this Bf 109F was claimed as a victory by *Capt* Tarasov of the *VVS* during November 1943.

the dark green band that indicated a new role, one which was only marginally more hazardous than that which the *Geschwader* had previously faced in the Mediterranean, although the comparisons were not exact. Typically, I *Gruppe* proudly displayed Hans-Joachim Marseille's name on the Bf 109G-6 flown by the *Kapitan* of 3. *Staffel*, the late 'Star of Afrika's' old unit.

Irrespective of the aircraft they flew, all frontline *Jagdgeschwadern* received these recognition bands, which remained in vogue for the duration of the war. In some cases, they were the only clue as to unit identity, as the popular pre- and early war practice of applying emblems began to lose its appeal as the war turned against the Germans. This had more to do with the sheer number of new aircraft that any given *Gruppe* took on charge to replace losses, than any slackening of morale on the part of the pilots, although this was also a factor. There was little point in the ground crew painters spending hours carefully applying a pilot's personal markings, only for that aircraft to be lost on the next sortie. As was quite understandable, given the circumstances under which they flew and fought, the top *Experten* got through multiple aircraft if their combat career lasted any length of time.

'Hubs' Mutterich and 'Jacky' Pohs, two of the *Experten* of JG 54 pose by the tail of the former's Bf 109G following the award of the *Ritterkreuz* to both pilots.

Hptm Heino Greisert, *Gruppenkommandeur* of 11./JG 2, in front of his aircraft marked with a double chevron at Abbeville in 1941. *Lt* Behrend is on the right. (*Crow*)

There was some small compensation for the Germans opposing the 8th AAF's P-51Bs in the early months of 1944 in that the Mustang experienced its share of teething troubles, not the least of which included the malfunctioning of the electrical system which could cause gun stoppages. Some American pilots lost all four guns during dogfights, a gratifying situation for the hapless *Jagdflieger* who might have been pursued by a Mustang at the time!

Depending (as always) on the calibre of the

The prototype Bf 109G-12 two-seater (CJ-MG) was a modified G-5 airframe.

respective pilots and the tactical situation, the Bf 109G was significantly outclassed by the P-51B, although it did not concede every point. Both fighters could climb at around the same rate and they were closely matched in rate of roll, but the Mustang could turn tighter than the Bf 109 and it could also outdive it by a considerable margin, either or both the latter factors often proving decisive.

It should be remembered however, that such comparative performance assessments were usually based on Allied figures amassed by RAF pilots of similar ability under ideal conditions. Transfer those figures to a cloudy day over the continent and replace the test pilot with a couple of inexperienced service pilots – and the demise of the Bf 109 at the hands of the P-51B was not quite so certain.

Apart from technical improvements, both sides adopted very cheap aids to combat efficiency. For the Bf 109, this was a black and white spiral design on the spinner of each operational machine. Creating a surprisingly effective optical illusion to throw bomber gunners off their aim, this was a welcome addition in a combat arena where every type of help was needed. Paint also helped each side identify friend from foe. The Bf 109 could easily be mistaken for the P-51B in the mêlée of combat, with occasionally fatal consequences and it cost little for the Americans to apply white wing and tail bands to their olive drab machines in order to decrease the chance of mistaken identity.

The task of countering the Mustang effectively absorbed the German fighter force in many hours of analysis. Much was already known about this remarkable American fighter's performance and weaponry, as relevant data had been available to front line pilots for some time. Rechlin had the opportunity to update this during the first week of June 1944 when the first intact P-51B became available for evaluation. Test flights more or less confirmed what many a *Jagdflieger* already knew – that here was a fighter almost without peer in the world. The Mustang test data was put to good use and a number of German pilots got to fly this and other examples which subsequently fell on the 'wrong side of the line'.

Numerous ways to combat the bombers effectively were explored: in the spring of 1944, *Inspektor der Jagdflieger* Adolf Galland put forward an idea to *Maj* Walter Dahl. Galland

In an attempt to improve the Bf 109's ground angle for carriage of an SC 500 bomb, an extra wheel was tested on Bf 109G-2/R1 (BD+GC), which was also flown at maximum external weight. *(via Robertson)*

envisaged a crack fighter unit created specifically for the interception of heavy bombers. Designated *Jagdgeschwader z.b.V.* (special duty wing), the unit was commanded by Dahl and based at Nuremburg-Ansbach. It comprised five existing *Jagdgruppen* – II./JG 3, I./JG 5, II./JG 27, II./JG 53 and III./JG 54, all equipped with the Bf 109G. A sound enough idea, JG z.b.V. was to exist barely more than one month, as after the Allied invasion, its *Stab* became that of JG 300 and the component *Gruppen* reverted to their own units.

There was with the advent of the P-51 in the 8th AAF, the additional hazard – even though this had been something the *Luftwaffe* had had to live with since 1941 – of enemy fighters strafing its aerodromes. This the American escorts began to do on a regular basis. To counter this, the Americans were offered 'soft' targets in the form of obsolete bombers dispersed on well camouflaged aerodromes heavily defended by *Flak*. Bombers, their usefulness waning with every passing month, could be sacrificed, especially if the enemy lost a significant number of fighters while attempting to destroy them. Nevertheless, some forward airfields became completely untenable because of marauding Allied fighters.

As the clock moved towards late May 1944, the German forces in France were subjected to a devastating series of attacks that disrupted rail, road and waterway communications, knocked out radar and radio stations and denied the *Luftwaffe* the use of even more airfields. The fact that the Allies had to blanket all coastal regions rather than specific areas in order not to reveal details of their vital invasion plan, could not entirely mask the fact that something big was brewing . . .

Invasion front

When the Allies landed in Normandy in June 1944, 12 *Gruppen* equipped with the Bf 109G (out of a total of 23) were hurriedly transferred to France to offer some resistance to the new threat. They were: III./JG 1, II./JG 2, II and III./JG 3, I and II./JG 5, II./JG 11, I., II. and IV./IJG 27, II./JG 53 and I./JG 301. But with units well below strength and mustering a combined total of just 289 serviceable aircraft, these could hardly expect to be very effective – or to survive long. Air combat and relentless attacks on *Luftwaffe* fighter bases severely depleted these emaciated units although the delivery of replacement fighters rarely faltered. And as long as they had the basic tools and facilities, the *Luftwaffe* groundcrews always managed to provide a handful of fighters for each day's sorties. What nobody could now do was to issue front line units with large numbers of well trained pilots.

Dire though the reality of Allied troops and tanks advancing in Normandy was for the Germans, the invasion area with its plethora of targets was more the province of *Jabo* attack rather than fighters, which suffered appallingly at the hands of the RAF and USAAF. Each Allied air force now fielded fighters which were equal – if not decidedly superior – in most important respects to the Bf 109 and Fw 190: the British had the Spitfire IX/XVI plus the Griffon-engined Mk XII, the Typhoon and Mustang III, the Americans the P-38J, P-47D and P-51B. As well as these highly capable single-seaters the *Jagdfliegern* would additionally be sent aloft to intercept a range of mediums such as the B-26 Marauder, A-20 Havoc, B-25 Mitchell and the Mosquito, none of which could be regarded as exactly easy meat!

Mosquitos had long been a thorn in the side of the *Luftwaffe* and two day fighter *Jagdgruppen* (JGr 10) had been formed in July 1943 with the primary object of intercepting them. Using Bf 109G-6s fitted with GM engine boost and with airframes highly polished to gain a few extra mph, anti-Mosquito sorties were flown on a number of occasions but neither unit achieved much success. However, the increasingly

A pressurised Bf 109G-1, as evidenced by the small air compressor intake above the supercharger air intake and solid vertical head armour to seal off the rear of the cockpit hood, almost certainly built at Regensburg. (D Howley)

frequent incursions by Mosquito bombers and intruders' raids would absorb additional *Luftwaffe* resources.

Fast reconnaissance

While the day fighters were finding increasing pressure on the invasion front, the *Luftwaffe's* reconnaissance *Gruppen* continued to carry out their little-publicised but vital role. Among the variety of aircraft flown by these units, were many Bf 109s which partially or fully equipped them, depending on their role. To cover Allied activity in the early weeks of the invasion, *NAGr (Nahaufklarungsgeschwader)* – short range army tactical reconnaissance group – 12 and 13 flew an intense series of high speed photographic sorties. Both had then been equipped with the Bf 109G-8, similar to the G-6 fighter but with fuselage frames 5 and 6 strengthened to accommodate cameras, these being either two Rb 12.5/7 x 9 or one Rb 32/7 x 9. Twin ports for the lenses were set into the bottom of the fuselage and early production G-8s also had a *Robot II* camera in the leading edge of the port wing, although this latter was not a success and was soon removed.

The photo reconnaissance G-8s, of which there were 167 built, served alongside the G-8/R5, which differed only in the type of radio equipment (FuG 16 ZS instead of FuG 17) and 739 of the latter were completed. Although the fuselage mounted guns were retained, some G-8/R5s had their engine-mounted cannon removed to save weight. In addition, the aircraft was wired to accept the gondola-mounted MG 151/20 cannon but it is unlikely that great use was made of these weapons. As was typical of the entire 109 series, modifications within this relatively small production batch applied, including GM-1 engine boost, ETC 500 bomb racks and the *Erla Haube*.

Hurling themselves against the mighty Allied juggernaut which they were clearly unable even to check, let alone halt, the *Jagdwaffe* was pulled back from France, having suffered substantial

Pleasing view of a Bf 109F-4 flown by the Technical Officer of *JG* 3 dispersed in a pasture. Despite its weak landing gear the Bf 109 operated quite successfully from basic grass and scrub landing strips most of the time.

casualties. Every interceptor was now more than ever needed in the *Reich*, for the destruction of heavy bombers had become, despite the increasing youthfulness and inexperience of the pilots, the primary reason for maintaining a fighter force.

More than once the very existence of the *Jagdwaffe* was called into question by Hitler who even went so far as to suggest that its personnel could be more useful fighting as ground troops. Defence was anathema to the *Führer*, who constantly demanded retribution in the form of increased *Jabo* attacks, long after these had become feasible. And fighter leaders were aghast at the simplistic view Hitler sometimes took of the 'poor' performance achieved by the *Jagdwaffe*, equating numbers of fighters to numbers of enemy bombers and strongly implying that if 60 interceptors took off, 60 bombers should automatically be shot down! Göring, his influence within Hitler's inner circle in decline, did not offer any encouragement by his remarks that openly accused the fighter pilots of cowardice and showed not the slightest appreciation of the problems they faced.

To help the novice pilots overcome their initial apprehension at the enormous responsibility they were handed, the German air ministry issued a comprehensive set of instructions, based on numerous combat reports, on how best to deal with the American heavies. They highlighted the weakest points of the B-17 and B-24, the best (and worst) direction of attack and the ranges over which the fighters' weapons were effective. This last was very important because accumulated data had shown that many pilots consistently under-estimated the range of their targets and opened fire too early. With only a limited number of rounds for his guns, the pilot who made such a mistake could not always expect to correct it before running out of ammunition. These investigations showed that on average a pilot hit the target with only two per cent of his expended ammunition.

The handbooks urged continual practise and familiarisation with gunnery techniques and stressed seemingly basic steps such as remembering to charge the guns. Such actions could easily be overlooked when the adrenalin flow

Small though it was, the Bf 109's cockpit (a G-2 is shown) suited thousands of German and Axis pilots so well that many preferred it to any other aircraft. The *Revi* gunsight mounting dominated the instrument panel.

increased. Above all, pilots were exhorted to close right in (to 400m or less) to ensure fatal hits.

While what they stated was profound, the manuals tended to disregard the fact that while the fighter pilot was carefully selecting his target, checking the round counter for his guns and sighting his angle of attack, all hell was breaking loose around him. He may have been lucky in that only a few hundred guns in the bombers would be blazing away at him rather than enemy fighters, but the sky over Germany in 1944 was no place for quiet, considered judgement. If he lived long enough to develop them, the *Jagdflieger* would be carrying out all his actions by simple reflexes . . .

Despite the odds and the confusion – the risk of collision was a not insubstantial factor to be considered – engendered by hundreds of aircraft sharing a sector of the sky at any one time, the *Luftwaffe* maintained a system of ground-controlled interception for as long as it enjoyed some warning of a developing attack.

To hit the American bomber boxes hard during the all-important first pass, controllers would advise the Bf 109 *Staffeln* to position themselves at altitude, preferably up-sun. Lower down, the armoured Fw 190s went in to attack the bomber squadrons that comprised the lower boxes. Mixed formations of Bf 109s and Fw 190s would be briefed to draw the escort away from the bomber-killers and on many occasions, this ruse worked. The trouble was that the Americans began to send so many escort fighters that entire squadrons or groups could be assigned a set task without any need to deviate; the mere presence of a substantial escort could be enough to deter the attackers – the close escort Mustangs did not need to charge off after every German fighter that approached the bomber boxes but merely to call in a different section of P-51s to deal with the threat.

On the right-hand side, the Bf 109G's cockpit had a fuse box and supplementary instruments, seemingly a bit low down for comfort!

But such carefully planned parameters were subject to unexpected changes, particularly if the weather transpired to help the Germans. Broken cloud could very effectively hide a *Schwarm* of Bf 109s which could carry out a swift, deadly attack and retreat into the murk which shielded them from the now-alerted escort. It was also common knowledge throughout the *Jagdwaffe* that the initial attack, in force, was invariably the most effective as the pilots would form up in their allocated positions and could usually maintain station as they made the high speed run-in.

Bf 109G fuselages in final assembly, each close enough for engineers to work on two aircraft simultaneously. Completed fuselages were wheeled away to have the wings fitted. *(via Robertson)*

Afterwards, each pilot broke away to avoid the escort or defensive fire from the bombers. At that point, the carefully planned formation would often have been depleted by at least one or two machines which were falling away, having been hit during the initial intercept. Fighters were then rarely able to reform in greater than *Staffel* strength for a repeat attack.

Airborne controllers worked with their *Luftwaffe* colleagues on the ground to place fighters along the bombers' route and concentrate them as far as possible to attack in relays. Each part of Germany was divided into sectors referenced to gridded maps and the *Jagdfliegern* grew used to flying sorties into the same sectors, sometimes day after day and week after week, as the Americans sent their Fortresses and Liberators against numerous types of industrial plants, communications centres and airfields along a number of recognised flight paths, which could only be varied to a limited extent.

In addition to literally tons of American bomber wreckage that had fallen on German-controlled territory since mid-1942, the *Luftwaffe* had acquired flyable examples of the two principal types for extensive evaluation. German pilots also made a point of examining shot-down bombers with the object of further understanding their most vulnerable areas. Neither the B-17F nor B-24D had been fitted with powered front turrets and the *Jagdfliegern* set much store by the head-on attack which was effective in that cannon and machine gun fire killed or incapacitated the flight crew and knocked out engines. Marginally more hazardous a practice when the frontal armament of the Fortress and Liberator was improved (in the B-17G and B-24H respectively), flying straight at a formation was not recommended for pilots of a nervous disposition!

At a target-closure speed of 600 mph plus, the fighter pilot had scant seconds to aim and fire before breaking over or under the bombers. A diving or climbing exit in either direction risked exposure to multiple guns but for a relatively short time; frontal attacks proved very effective and such were responsible for the demise of numerous bombers, but they tended to be dropped in favour of more conventional intercept angles towards the end of the war because many German pilots, particularly those with little experience, believed them to be marginally safer.

As the Allies advanced inland from Normandy and the RAF heavy bomber offensive widened to take in a range of tactical targets in daylight in concert with the USAAF as distinct from its nocturnal pounding of German cities, so the need for interceptor fighters became even more acute. There had for some time been little action for crews of the *Kampfgeschwadern* based in the West and it was therefore decided to disband almost the entire bomber force and transfer the pilots and navigators to fighter training.

For much the same reason, few of the *Zerstörer Gruppen* survived much beyond this point, for these too were able to provide a nucleus of trained personnel for the new *Jagdgruppen* that were formed in August 1944. These mainly served to bring up to strength units that had not previously had their full complement of three or four *Gruppen*. Among them was JG 4, which had II, III and IV *Gruppen* added, II *Gruppe* being formed from the Ramm *Staffel* and IV *Gruppe* from a nucleus of personnel provided by JG 5. II./ZG 1 briefly became III./JG 76 before being renumbered as IV./JG 53. JG 4 was also among the completely new *Jagdgeschwadern* raised at this time, as was JG 6. Personnel for its first two *Gruppen* were provided by ZG 26, I./JG 5 being renumbered as III./JG 6.

Chapter 9
East–West Debacle

When the Soviet summer offensive began on 10 June 1944, the Germans were surprised that the Russians moved against their Northern positions rather than those on the Central Front, as expected. It was strongly suspected that this was a feint and within a fortnight, Russian intentions were clearer. With their tactical airpower causing widespread chaos among *Wehrmacht* infantry and the artillery units which were a primary threat to their tanks, the Red Army surged forward, aiming to break through the weakly-held bulge in the centre in several places. The *Jagdwaffe* was hard put to stem the tide: too few combat-hardened fighter pilots were available in any part of Russia.

The German front line then ran from the Baltic in the North through Estonia, snaked down to bisect the River Dnieper, bulged back into the area of the Pinsk Marshes and down to the Western bank of the Dneister. The Germans held Sevastopol but had lost Odessa on the Black Sea coast.

As far as fighter support was concerned, the new Soviet threat could only be contained by denuding other fronts which could ill-afford to spare pilots. Units did nevertheless transfer to the

Abandoned in the face of the Russian advance, this Bf 109G-14 awaits scrapping at an airfield in the Minsk-Bobruisk area in 1944.

East, there to find a situation that forced a series of base changes almost immediately to further disrupt daily operations. Four *Gruppen* of fighters arrived plus ground attack units, the latter being given the all-but impossible task of containing the advance of hundreds of tanks, self-propelled guns and troops. It simply could not be done on the required scale.

Nevertheless the *Jagdwaffe* made an orderly, 'fighting retreat' and put up as many sorties as possible while doing so, leaving only the minimum number of aircraft behind and destroying those that could not be evacuated. At new, sometimes hastily-prepared airfields, minor damage in landing or take off accidents could effectively write off fighters which there was no time to repair and at some locations, fuel and ammunition supplies could not be flown in in time. The experience of I./JG 51 was typical. Over a period of eight days, the unit moved back to Pinsk from Orscha, the base in the very centre of the front it had occupied on 23 June, via Dokudovo, Bojary, Baranovicki, Minsk and Pukhovichi, losing ten aircraft in the process. Many Russian airfields lacked hangars or other fixed installations, being merely large open areas of flat ground. During the retreat, German units returned to familiar surroundings more often than not.

By July the Soviet advance had gathered a momentum that was seemingly unstoppable and in August an additional offensive in the South was opened. Making excellent progress, Red Army troops had crossed into part of East Prussia during the last week of the month. There followed further setbacks for the Germans, leading to the loss of much Eastern European territory as a result of diplomatic rather than military moves.

Romania, doubtful about the ability of the Germans to hold the Russians, promptly staged a *coup d'état* and declared war on their former ally. Bulgaria, an unwilling partner of the Axis, claimed neutral status and attempted to make peace with Stalin. After an anxious four days, an armistice was signed on 9 September. Germany was thus denied the support of two Axis states in less than one month and then Finland, having won back the territory conceded to the Russians in the Winter War, stopped fighting and began denying the Germans port facilities in the Baltic.

A perceptible slackening of pace brought the Germans a breathing space in October, the Red

A Bf 109G-6/R flown by the Czech-manned 13./JG 52 *(Milan Krajci)*

Army's incredible advance having dangerously stretched its supply lines. Hard fighting had exhausted the troops and even with their massive forces and substantial reserves, the Russians inevitably had to slow the pace as the autumn brought less favourable conditions.

Luftflotte 6 now held the key to German air operations on the Central Front. On a handful of bases in Poland and East Prussia, it had about 1,000 aircraft of all types, all of them heavily outnumbered over the combat zones and having to take heed of warnings to conserve fuel. Even worse off in terms of aircraft was *Luftflotte* 4 based in Hungary and Yugoslavia. Down to 200 combat aircraft, it too awaited the next Russian assault with little hope of being able to achieve much to materially assist the *Wehrmacht* to hold or to regain lost ground.

The Soviet offensive had isolated *Luftflotten* 5 and 1. The former was now more than ever tasked with defending the far northern regions and attempting to fill the void left by Finland's withdrawal while the latter, cut off in the Courland peninsula, was by-passed by the main body of Russian troops. It was the weather that once again came to the aid of the Germans; having endured two winters in the East during which the elements appeared to side entirely with the Russians, they realised that this time, overcast skies and freezing temperatures created a slim lifeline for the forces fighting on both Eastern and Western Fronts.

Ray of hope

Deteriorating weather enforced a lull in the fighting on both fronts as the winter of 1944/45 ushered in cloud-covered skies and a welcome respite from the attentions of the Allied air forces. In the West, the *Luftwaffe* reaped some benefit from a lull in the pounding of its synthetic oil refineries as the Allied heavy bomber offensive was switched temporarily from strategic targets to tactical ones in support of the land armies. Fuel supplies stabilised and rose and in September the aircraft factories delivered an all-time monthly record of 3,821 new machines to *Luftwaffe* depots. About four-fifths of the total was made up of Bf 109s (1,605) and Fw 190s (1,391), enabling 15 fighter *Gruppen* to fully re-equip.

The eleventh hour call for more fighters had been met: from a strength of 1,900 aircraft in early September the *Jagdwaffe* was expanded to 3,300 machines by November. Each day fighter *Gruppen* in the West had an extra fourth *Staffel* added, each with 10-15 aircraft, the pilots to fly them coming from the training schools or disbanded *Zerstörer* and bomber units. This gave each *Jagdgeschwader* a total of 16 *Staffeln* and a nominal strength of upwards of 160 aircraft.

The switching of resources to fighter construction at the expense of conventional bombers left many hundreds of the latter idle and various schemes were put forward to expend these in unconventional attacks on Allied ground targets. The Bf 109 and Fw 190 were to become an integral part of one such, the *Mistel* programme.

By mounting a fighter above the fuselage of a Ju 88 bomber on a trapeze arrangement of struts and rewiring the controls so that the Bf 109 or Fw 190 pilot 'flew' both aircraft, the *Luftwaffe* created a potentially formidable weapon. Control of both machines was surprisingly easy for the fighter pilot, who approached the vicinity of the target before releasing the bomber by firing a series of explosive bolts on the trapeze. After separation the fighter would fly back to base for a conventional landing, the Ju 88 meanwhile terminating its shallow dive into the target in a huge fireball. The nose of the Ju 88 was packed with a shaped explosive charge, the detonation of which together with the aircraft itself, offered the chance to pulverise many targets that would otherwise have required many tons of explosive if dropped by conventional means.

Kampfgeschwader 101 carried out the first operational *Mistel* sortie on 24 June 1944. Four combinations, escorted by Bf 109s, took off from St Dizier to attack invasion shipping off the Normandy coast. This and about nine other *Mistel* operations met with little success; when the Allied push inland drastically decreased the number of shipping targets a plan evolved to carry out a mass *Mistel* attack on the Royal Navy's anchorage at Scapa Flow. Development of the concept therefore continued and centred increasingly on an Fw 190 director rather than a Bf 109 although most of the early experimental flights were made with the latter. The Scapa Flow operation was never actually flown and the *Mistel* programme remained one of many basically sound military aviation projects that could only be partially developed by the Germans before the end.

It had long been realised that the precariousness of the German position would only be rectified by hundreds of single-seat fighters flown by able pilots, not by potentially superb but unproven weapons or curiosities such as the *Mistel*. Whether the fighters were powered by reciprocating engines or turbojets would make little difference, for pilot quality (and availability) was fast becoming the crucial factor in Germany's continued ability to wage an air war.

During October 1944, Messerschmitt initiated production of the penultimate main variant, the Bf 109G-10. As noted previously, this sub-type was produced out of sequence and followed the G-14 rather than preceeding it. It was a link with the G-series and K-series and was an attempt to bring older aircraft up to Bf 109K standard to offset production of new Bf 109K airframes.

As operational units began to receive the G-10 (most front line Bf 109 units eventually flew this variant), modifications were introduced and as had happened previously, the aimed-for standardisation could not be achieved. A primary factor in this was that engine availability and output in the autumn of 1944 could not be as reliable as in previous years due mainly to the disruption caused by bombing and a dwindling supply of

Erla-built Bf 109G-14 (Wr Nr 463141) 'Black 17' joined Fw 190D-9 (Wr Nr 211934) of II./*JG* 6 and a Ju 87D to surrender to the Allies at Furth on 8 May 1945. All three types were among the most active *Luftwaffe* aircraft during the last weeks of the war. (*Crow*)

A Bf 109K-4 with abbreviated Wr Nr '6316' abandoned at Kassel in 1945. (*Crow*)

raw materials. Consequently, the first Bf 109G-10s left the assembly lines powered not by the intended DB 605 D but the DB 605/AS. Known as the Bf 109G-10/AS, this sub-type was otherwise identical to the G-14/AS.

When the DB 605 D was installed in the Bf 109G, the powerplant was known as the DB 605 DM to identify the addition of MW 50 boost. Some necessary changes were made to the aircraft's engine cowling and a noticeably deeper oil cooler fairing was fitted under the nose. Aircraft in different production batches varied in other ways including the type of tailwheel fitted (long or short stroke, retractable or fixed); some machines had the larger 660 x 190 mainwheels and either the small or large vertical tail surfaces.

Production of the G-10 coincided with the output of the first of the Bf 109K series, which was revised before quantity production had been started. Instead of producing a number of early variants and progressively modifying each succeeding one so that a performance peak was only achieved gradually, Messerschmitt broke with tradition, scrapped the Bf 109K-0 to K-3 and concentrated resources on building the K-4, aiming once more for a high degree of standardisation.

Powered by the DB 605D engine, the Bf 109K-4s were new-build aircraft and not, as with the similar G-10, conversions of older airframes. A vast production programme for what was to be the final Bf 109 variant was constantly revised and the actual number of aircraft delivered was a fraction of the initially planned figure of 12,700. Interestingly, Messerschmitt did not envisage K-series production being completed until March 1946.

It was mid-October 1944 before the first K-4s reached front line *Gruppen* in the West and 534 aircraft had been delivered by the end of November. The K-4 therefore entered service before the G-10 and by the end of the war, Messerschmitt's parent plant had completed well over 1,000 airframes. Erla did initiate production although it is believed that only a very few K-4s were actually delivered by this concern. The K-4 also saw service on the Eastern front but probably not before 1945.

Externally similar to some of the Bf 109G-10s, the K-4 had the wide-bladed VDM propeller and the deeper Fo 987 oil cooler as well as the two bulged side fairings enclosing the larger crankcase of the DB 605D, or DB 605DM as it was known, MW-50 power boost being standard equipment. In addition, the Bf 109K-4 re-introduced the efficient radiator coolant cut-off valves which had been retro-fitted to a number of F-series machines in 1941.

The *Erla Haube* was fitted as standard and many aircraft had the radio aerial mast deleted. This was due to the fact that the mast could easily be damaged when the hood to which it was attached was opened too far. It was a relatively simple maintenance procedure to dispense with the mast and attach the aerial to a hole in the fuselage spine, feed it through the D/F loop and affix it to the fin. The K-4's fin and rudder was the larger unit, and usually had a Flettner tab and two fixed tabs – although there were examples without the fixed tabs.

Short and long tailwheel types were fitted although all K-4s were to have been fitted with the longer, retractable unit designed to be enclosed by twin doors. Most K-4s also had the larger mainwheel hubs and tyres introduced on

All but complete, with only the canopy smashed. 'Black 21' a Bf 109K-4, awaits its fate after the surrender. (*Olmsted*)

the G-10, with small doors enclosing the mainwheels. The doors were often removed in front-line service.

The engine-mounted MK 108 cannon had been standardised for the K-4, as had two MG 131 machine guns in the forward fuselage. Surprisingly, considering the kind of opposition the Bf 109K-4 was destined to face in service, there was no standard provision for underwing cannon but these guns could be fitted as a *Rüstsatze* in the K-4/R4, few of which were actually completed. The Bf 109K was therefore destined to operate with a fairly modest armament until the end of hostilities, as indeed had the entire Bf 109 series.

Component *Gruppen* of all the first line *Jagdgeschwadern* except JG 5, 6 and 54 were at least partially equipped with the Bf 109K-4 as ample numbers of G-6s, G-10s and G-14s remained in service, backed by substantial stocks in *Luftwaffe* depots and at the factories. It was often the case that to save time, personnel would collect new aircraft from increasingly full depots rather than carry out time-consuming field repairs to machines damaged in combat or training. Full Bf 109K-4 equipment was achieved by III./JG 3, III./JG 26, III./JG 53 and III./JG 77.

Despite the fact that the later Gs and the K represented the most powerful Bf 109s to date, the type's armament remained unchanged at the maximum of three cannon (two as optional fittings) and two machine guns. Reflecting what had been required since the appearance of the Bf 109F, Messerschmitt proposed reintroducing wing guns but only in a very late variant, the Bf 109K-6 'heavy fighter'. This, with two MK 108 cannon in the wings outboard of the main undercarriage wells, could ostensibly have been a far more formidable fighter – certainly in the anti-bomber role – than any other Bf 109 hitherto built. But the K-6 was not produced in any numbers and there is some doubt that it was ever issued to an operational unit, although one example was under test at Rechlin as early as the autumn of 1944.

Having managed partially to revitalise itself, the *Jagdwaffe* was able to put 490 fighters into the air to attack a force of 1,000 USAAF heavies and its fighter escort on 2 November. These bombers were out to wreck the Leuna-Merseburg oil refineries, a target which was just one of a long list of German centres of vital fuel and petroleum. Destruction of such plants would speed the total demise of the *Luftwaffe* as surely as the cancer of pilot losses and the American raids were consequently contested as far as possible.

Spirited attacks by *Sturmgruppe* (assault group) Fw 190s destroyed 22 B-17s but the P-51 escort had quickly wiped out this success by shooting down 31 of the German fighters. In attempting to even the odds, JG 27's Bf 109s gamely waded into the Mustangs but lost 27 pilots. Overall, 2 November was a terrible disaster for the German fighter force, which recorded 70 pilots killed and 28 wounded, with 120 aircraft lost.

Forty B-17s had been lost to 8th Air Force strength, plus 16 fighters. In both air forces, lost or damaged aircraft would be replaced relatively quickly but the Americans had the advantage in that their replacement pilots and crews would have had excellent training and probably some useful combat experience so that the leadership qualities and flying skills of those who had gone down would not create as, in the *Luftwaffe*, a void that was extremely hard to fill.

The 2 November debacle, the worst in the *Luftwaffe*'s history to date, had further adverse effects on the *Jagdwaffe* for it was the outcome of this series of air battles that all but convinced Hitler that the 'big blow', the proposed launching of multiple hundreds of fighters to all but annihilate a single enemy bomber force, might not result in the desired number of kills. His mind turned more to using the fighters in a ground attack operation, much to the chagrin of commanders such as Adolf Galland, who had slaved to marshal enough fighters for a really decisive operation against the bombers. Galland was not informed of Hitler's true intention, which was to launch a major ground offensive which would require substantial air support, rather than a large scale anti-bomber attack.

And it was all but impossible to preserve this force intact and separate from the daily round of operations; regular sorties still had to be flown by the available pilots and aircraft (although these latter could be replaced by new machines for as long as the factories were able to build them). Fighter operations for the rest of November 1944 brought more staggering losses to a force already reeling; on the 21st, 62 pilots were killed and wounded with 87 succumbing on the 26th and 51 on the 27th. A fighter ratio of as many as 30 to one in favour of the Allies was what the *Jagdwaffe* could now expect when planning the interception of a large scale bomber force.

Tactics were revised in the face of these daunting odds, although they were hardly very successful for the defenders. Bf 109s continued to undertake interception of the bomber boxes flying at the higher altitudes, insofar as the German reaction could still be neatly compartmentalised. A *Rotte* (often all that could be mustered) of fighters would dive out of the sun onto a Mustang formation with the object of making the Americans drop their tanks and thereby limit their radius of action before boring through to hit the bombers. This manoeuvre was however far from as effective as it had previously been; American fighter pilots were no longer obliged to nurse a fuel-starved fighter all the way to England, for there were plenty of diversionary airfields available on the continent.

The Bf 109s would open fire during their high speed dives, aiming to pick off one or two of the enemy but rarely now did the *Jagdfliegern* court disaster by staying around to dogfight with Mustangs, for the possibility of surviving such an encounter was very slim indeed. If possible the P-51s would be diverted long enough for a *SturmGruppe* (attack Gruppe) of Fw 190s to charge into the bomber boxes and pick off one or two, but it was an exceptional event if the US casualties reached double figures.

Much hope had been placed on the Me 262's superiority over conventional enemy fighters being able to redress the balance in terms of bomber interception but this was generally dashed. Owing to numerous problems, not the least of which were technical faults traceable to hurried development, manufacturing problems and sub-standard raw materials, the jet fighter threat largely failed to materialise and the number of victories claimed by pilots could not alter the balance. Although it was potentially excellent, the Me 262 actually became something of a liability for the *Jagdwaffe*, as conventional aircraft strove to protect it from marauding Allied fighters, a task they were often very hard put to achieve without sustaining substantial casualties.

Experimental 109s

With the Bf 109 having already gained a reputation as a high altitude fighter in standard form, Messerschmitt had initiated work to better the service ceiling of the Bf 109F (39,370 ft for the F-3) by a substantial margin. The *RLM's Spezial Höhenjäger* (Extreme Altitude Fighter) programme of 1942 called for such an aircraft with a secondary reconnaissance capability, but to save time, utilising as many standard components as possible. Both Messerschmitt and Focke-Wulf were asked to submit designs.

A standard Bf 109F airframe was consequently fitted with an additional 6 ft 6.5 in wing centre section and flight-tested at Rechlin as the Bf 109H V1. This machine was destroyed in an air raid in August 1944, by which time a small batch of Bf 109H-0 pre-production aircraft had been completed, these being followed by a number of Bf 109H-1s. Powerplant was the 1,300 hp DB 601E and the armament a single MK 108 cannon and two MG 17 machine guns.

Flight tests of the Bf 109H-0 revealed a typical service ceiling of 47,500 ft, this being some 6,000 feet more than attainable by a standard service Bf 109K fighter, but slightly inferior to that of the extended-span Ta 152H-1, which was capable of 48,550 ft with GM-1 boost. As the Bf 109H would have been a direct competitor to the Ta 152 (and probably not as capable an aircraft in service), the *RLM* abandoned any further development of the Messerschmitt design.

Before that however, some if not all the Bf 109H-1s were delivered to the *Luftwaffe* reconnaissance unit 3.(F)/121 based at Bernay in April 1944 and it would indeed be interesting to know how these machines fared in service. Available records do not indicate whether the Bf 109Hs were operated as standard armed fighters or configured as unarmed reconnaissance aircraft carrying a single Rb 50/30 or Rb 75/30 camera, as proposed by Messerschmitt for this latter version.

The modest dimensions of the Bf 109 did not lend themselves to overmuch development – anything revolutionary would have inevitably required a redesign of the fuselage, such as was done with the Me 209 record-breaking aircraft of 1937. Willy Messerschmitt himself was however, confident that the basic 209 concept could have been developed into a new fighter for series production and despite the Me 309 failing to meet its projected performance, design work on a Bf 109 successor continued.

This was given impetus by the *RLM* following a meeting on 5 July 1943, which outlined to Messerschmitt and Focke-Wulf a *Schnellösung* (quick solution) to the high altitude fighter requirement, the specification for which both companies had found difficult to meet. In Messerschmitt's case, the result was the Me 209-II, which offered considerable potential. By utilising some 65 per cent of Bf 109G components mated to a new wing spanning nearly 36 feet, with a tall, narrow chord vertical tail and powered by a 1,750 hp DB 603A-1 engine with an annular radiator, the new fighter bore a superficial resemblence to the Fw 190D/ Ta 152 series which it would have complemented in *Luftwaffe* service.

Flying for the first time on 3 November 1943, the Me 209 V5 (this *Versuchs* (experimental) number was allocated to avoid confusion with the quartet of Me 209s, built as the V1 to V4) redressed the main shortcomings of the Bf 109, particularly the narrow track undercarriage. Installation of a Jumo 213E engine for the Me 209A-2 production variant did not adversely affect performance, the aircraft being capable of a top speed of 410 mph and a service ceiling of 42,650 feet.

Flight testing continued with two further examples, the second one flying for the first time on 22 December 1943. The third aircraft, which eventually took the designation Me 209A-2, began flight tests in May 1944, but the *RLM* decided that disruption of the Bf 109 production line at that point of the war would be disadvantageous, particularly as the higher performance Fw 190D-9 was then preparing to enter service. The Me 209 fighter programme was therefore cancelled although Messerschmitt completed a fourth airframe as the Me 209H V1 with new inner wing bays similar to those of the Bf 109H, to extend the span to 39 feet. This aircraft began flying in June 1944 but its ultimate fate is unknown.

Among other projects associated with the Bf 109 in more or less its original configuration was the 109Z *Zwilling* (Siamese Twin), whereby two fuselages were joined by incorporating a new wing centre-section and a parallel-chord tailplane. Intended for the *Zerstörer*/fast bomber role, the first Bf 109Z utilised two Bf 109F fuselages each fitted with double undercarriage legs, the wing centre section incorporating a pair of radiators and provision for a single MK 103 cannon. This machine was all but complete before it was damaged in an air raid and no flight testing apparently took place before the project was abandoned in 1944.

Less ambitious Bf 109 modifications included the fitting of a V-tail tested on a Bf 109F. This was successful and indicated that such a modification might improve stability although disruption of the production lines was seen as detrimental and the idea was not proceeded with. An interesting idea whereby the Bf 109 was adapted to carry an additional undercarriage leg was also tested. Positioned at mid-fuselage point this jettisonable leg raised the tailplane clear of the ground and gave the aircraft an almost level ground angle to enable the carriage of a single 500 kg (1,102 lb) bomb. As the Bf 109G-2/R1 long range fighter bomber, the aircraft also carried two 300 ltr (66 Imp gal) underwing drop tanks. Although successful in flight tests, no production was undertaken.

Chapter 10
Italian Swansong

As Hitler's principal partner in the Axis Pact, Mussolini enjoyed some priority on German aircraft reinforcements for the *Regia Aeronautica*, should the war situation demand it. In terms of modern fighters such a need had become acute by early 1943, for while Italy's modern fighters, the Macchi C 202/205, Reggiane 2002 and Fiat G 55 series were excellent aircraft, they were few in number and too modestly armed to combat contemporary Allied types without a prohibitive rate of attrition. Consequently, Italy received a substantial number of Bf 109s within a total of more than 700 German combat aircraft transferred during the remaining war years.

Germany initially furnished some 50 Bf 109s drawn from the inventories of the Italian-based *Luftwaffe Jagdgeschwadern* which then consisted of *Gruppen* of JG 27, 53 and 77. Personnel of two *Grupo Autonomo Caccia Terrestre* (3o and 150o) with a total of six *Squadriglia* (*Sq*) were trained on the Bf 109, starting in April with men from the latter unit based at San Pietro.

Although equipped only with 12 decidedly 'war weary' Bf 109F-4s, it was hard for some Italian pilots to suppress the urge to throw their mounts all over the sky, just the way they had been taught to fly their own fighters. Old habits die hard and *Regia Aeronautica* doctrine had traditionally encouraged them to exploit the inherent aerobatic qualities of Italian fighters in combat. The Bf 109 was a totally different concept.

Unsurprisingly, the Italians had their share of accidents with the Bf 109 although most pilots were far from novices. They found the *Friedrich's* light aileron control to be demanding during landing and as always, that delicate undercarriage had to be treated with great respect.

150o Gruppo initiated Bf 109 combat operations with 22 109G-4s and -6s formerly operated by JG 53, in April 1943. Initially sorties were flown to gather intelligence on enemy shipping movements and to escort ASR seaplanes. Pilots of 3o Grupo commenced Bf 109 conversion training on 23 May, the unit having received 23 G-6s at Comiso. Not all of these machines were serviceable with the result that full conversion was only completed slowly. Neither did the Italians now have the benefit of German instructors.

The Italians were hardly short of targets during the aerial blockade of Pantelleria, their Bf 109s clashed with enemy fighters for the first time. On 28 May 365a *Sq* intercepted bombers escorted by Spitfires but no claims were made. A period of intense activity over Pantellaria followed and on 10 June both sides filed air victory claims.

In company with the C 202s of 53 *Stormo*, 150o Grupo intercepted a formation of 340th Group B-25s covered by 31st FG Spitfires. The Macchis went after the Mitchells while the Bf 109s took on the escort. *Ten* (Tenente) Ugo Drago, CO of 363a *Sq* claimed two Spitfires before he was himself nailed by another and forced to bale out. *Ten* Chiale also claimed one victory although the Americans claimed five. That same day, pilots of 150o Gr collected seven new Bf 109G-6s allocated by the *Luftwaffe*.

On the 13th the Italians were able to celebrate their first Bf 109 victory when *Serg Magg* Carlo Cavagliano of 5a *Sq* joined II./JG 53 on a bomber interception. The enemy was heading for Comiso, home to the Italian and German units and it was marginally safer to be airborne than under the bomb carpet. With the bombers in sight, a Spitfire was spotted pursuing a German Bf 109. The RAF aircraft was chased by *Serg Magg* Cavagliano who opened fire. Hit in the engine, the Spitfire fell away and the pilot baled out.

By mid-June 3o and 150o Gr had fully converted to the Bf 109G-4 and G-6. Under rarely possible ideal conditions the lighter G-4 was used for fighter combat while the more heavily-armed G-6 was reserved for intercepting bombers. The number of Bf 109s delivered to the two units steadily increased so that by late June for example, 3_0 *Gr* had 34 aircraft available. The first half of June had seen the Italian Bf 109s score 23 kills in 323 sorties, for the loss of 13 aircraft. *Ten* Giovanni dell' Innocenti of 150_0 *Gr* led the list of successful pilots with 12 victories.

A slight drawback with the Bf 109s supplied to the *Regia Aeronautica* was the lack of an R/T set. German fighter units in Sicily were directed by two radars which transmitted data to be relayed to pilots via the FuG 25 IFF set. There had not been time to convert Italian pilots to use this system and therefore the sets were removed from most of the Bf 109s supplied.

When the Allied storm broke over Sicily, Axis losses of aircraft on airfields subject to repeated bombing began to rise alarmingly, and the low flying enemy *Jabo* attacks were often as deadly. Repairing damaged or wrecked aircraft with few spares and equipment created a logistical nightmare.

On 18 June *Sotto Tenente* (Junior Lieutenant) Antonio Camaioni of 363a *Sq* flying from Sciacca, became involved in a violent air battle with USAAF Lightnings protecting a formation of A-20s of the 17th and 47th Bomb Groups. Two 1st FG P-38s fell to Camaioni's guns before he had to retire, his Bf 109 severely damaged. With about 100 fighters based in Sicily, the Italians managed to shoot down 53 enemy aircraft for the loss of 15 in the air but an illustration of how ground attacks could be more destructive than air combat is shown by the loss of 24 Bf 109s on the ground between 26 June and 4 July.

There was no shortage of courage on either side and the Axis (invariably heavily outnumbered) continued to challenge Allied formations. All too often the results achieved were mere pin-pricks but seasoned Axis pilots could occasionally exploit the situation to their advantage. On 3 July the Baltimores of No 21 Sqn SAAF briefed to attack Trapani-Milo airport, covered by P-40s of the 33rd and 324th Fighter Groups were intercepted by the Bf 109Gs of 150_0 *Gr*. The Baltimore crews grimly noted how the Italians drew off the inexperienced American fighters, leaving them all but unprotected. Four Baltimores were shot down by Messerschmitts of 364a *Sq*, pilots of which put holes in several more of the Martin bombers before departing. Seven P-40s were claimed shot down in a number of dogfights, two by *S Ten* Camaioni. Only one Bf 109 was lost, that of Giovanni dell' Innocenti, one of 150_0 *Gr*'s high scorers.

Further combat between 150_0 *Gr* and US P-40s were successful for the Axis side and the score of medium bombers also gradually increased; but the pace was hectic. Three alerts before midday became commonplace, with further interception sorties being posted in the afternoon and early evening. Bombers in two waves hit San Pietro on 6 July, destroying 13 Bf 109s on the ground. With Italy's political position about to change radically, the Bf 109 units were increasingly used for reconnaissance sorties.

On 3 July Operation HUSKY, the Allied invasion of Sicily began. Two Bf 109s of 150_0 *Gr* were up early from Sciacca to report shipping movements. Both were intercepted by Spitfires, one being shot down and one damaged. Better results were obtained from an afternoon interception by 150_0 *Grupo*'s Bf 109s of five P-40s of the 324th Group. Three Warhawks were shot down by *M.Ilo* Walter Bertocci although his own aircraft returned home with a wing shredded by machine gun fire.

As things transpired, that was the last combat for 150_0 *Gr* over Sicily. By 9 July this unit's complement had dwindled to six serviceable aircraft out of 21 Bf 109G-4s and G-6s based at Sciacca. Over at Comiso, 3_0 *Gr* had not a single serviceable aircraft left out of a complement of 21 G-6s. The following day, the first Allied troops came ashore.

Hamstrung by wrecked airfields, the fighter units in Sicily were pulled back to Italy. Repeated appeals to *Oberbefehlhaber Sud* for replacement Bf 109s (300 Bf 109Gs were requested on 12 July) could not be met and 3_0 and 150_0 *Grupi* personnel abandoned even partially-damaged aircraft to the enemy before leaving. Three 150_0 *Gr* Gustavs remained just about flyable at Sciacca and on 12 July, these were hurriedly patched up and flown to Palermo to be reunited with the rest of the unit.

Three Bf 109Gs battered from the Sicilian fighting, hardly constituted a very formidable force and at this point both 150o and 3o Gr – the latter having lost all its aircraft – began a period of relatively low-key activity while the future of Italy's war effort was decided. Despite the 3 September armistice with the Allies, Hitler's avowed intention to occupy the country and continue the fight presented many Italians with an agonising choice.

With the armistice, the *Regia Aeronautica* ceased to exist and air force personnel experienced a confusing and uncertain period for a few weeks until 27 October when a pro-fascist air arm, the *Aeronautica Nazionale Repubblicana (ANR)* came into existence. In effect the Italian air force had split into two as a nominal North (German-held) and South (Allied-held) division. Those individuals who sided with the Allied cause became part of the Co-Belligerent Air Force.

To many, the decision to fight on alongside their erstwhile allies was merely an extension of what they had been trained for and believed in – a national air arm under their leader, Benito Mussolini. Men who had known little but the fascist doctrine did not care to suddenly switch their allegiance, despite the fact that if the war situation continued to isolate and strangle Italy, siding with the Germans was all but a lost cause.

An initial fighter organisation, I *Grupo* Caccia, was formed in November 1943 but it was 1944 before the formation of II₀ *Grupo* was authorised. Italian units were now organised on *Luftwaffe* lines which differed markedly from those of the former *Regia Aeronautica*. Ground support and intelligence gathered by the radar network in northern Italy was provided to the Italian units by *Jagdführer Oberitalien* (Fighter Leader Italy).

By the spring of 1944 arrangements had been made for a cadre of *ANR* officers and NCOs to train in Germany on the Bf 109, the first contingent leaving on 29 April. *General* Adolf Galland visited the trainees at Leignitz on the 30th. For some pilots, the transition would be quite smooth as they had either flown the Bf 109 or models of the later Macchis or Fiats. Training accidents were few and it was decided to convert all three *Squadriglia* of II *Grupo* to the Bf 109 while concentrating Italian fighters in I *Grupo*.

It was agreed that the Italians would take over 43 Bf 109G-6/R6s which had been ferried to Ferrara and Piacenza airfields by pilots of I./JG 53 and II./JG 77. A further training period was now initiated, primarily for those pilots of 3a *Sq* who had not flown the Bf 109, at Cascina Vaga. A Bf 109G-12 trainer was requested to speed conversion and, to make good any shortfall through accident or unserviceability, a number of *Centauros* were retained. M Ilo Artidoro Galetti, who mastered the Bf 109 particularly well, had completed his training by June.

The syllabus included flying the standard *Luftwaffe Schwarm* formation, the interpretation and use of data gathered by *Jafü Oberitalien* radar, general tactics and gunnery. II *Gr* pilots, perhaps nostalgic for the lighter, more manoeuvrable Italian fighters they had traditionally flown, removed the gondola-mounted 20-mm cannon from their *Gustavs*. Considered vital for the destruction of well-protected Allied bombers, these additional guns did little to enhance the G-6's performance and their removal did indeed make the aircraft much more manoeuvrable, if less effective as a fighter.

After the almost inevitable training accidents the *ANR*'s Bf 109 *Grupo* was considered to be operational and 1a and 2a *Squadriglia* took up station at Cascina Vaga on the south side of the Po River, this being the most southerly of the bases used by the Italian Bf 109 units. On 22 June six Bf 109G-6s drawn from each *Squadriglia* recorded the *ANR*'s first Bf 109 combat sorties to intercept B-24s, without any positive results.

The first victories for *ANR* Bf 109s came on 24 June. *Ten* Ugo Drago, *Ten* Raffaele Valenzano and *Serg Magg* Carlo Cavagliano of 1a *Sq* and 3a *Sq* attacked *Gr* II/3 'Dauphine' P-47s bombing and strafing the railway near Genoa. The Italians waded in and in the ensuing dogfights, *Ten* Drago and *Serg Magg* Cavagliano accounted for one P-47 apiece.

Combat on 29 June resulted in the USAAF officially reporting the existence of Italian-operated Bf 109s when Lt Mannon of the 64th FS, 57th FG opened fire on an *ANR* aircraft during an air battle over Forli. 1a and 2a *Sq* put up three *Gustavs* each accompanied by six G 55s and had shot down an A-20 before tangling with the

-47 and RAF Spitfire escort. One G 55 was destroyed.

Air actions through to the end of June resulted frustratingly modest victory claims and on occasions, non-substantiated 'kills' which pilots believed they had achieved in the heat of combat. July began on a dismal note when two more pilots succumbed to the odds during a fight with P-47s escorting the Italians' prime targets at this time, namely Douglas A-20s.

The Allied air forces now began what turned out to be a lengthy 'final' wearing down of the German and Italian forces in northern Italy by initiating in July 1944, Operations MALLORY and MALLORY MAJOR, both aimed at stopping supplies moving across the River Po. Rail bridges were prime targets, as was the entire Italian rail network in the region able to maintain the flow of supplies from Germany via Austria.

With precious few fighters available for bomber interception, *ANR* pilots realised how soon their ranks could be decimated, should they be surprised by Allied escort fighters. A number of special fighter sections nicknamed *Occhio di Lince* (Lynx Eye) were therefore formed. Led by *Ten* Raffaele Valenzano the first such section was completed by *Ten* Camerani, *M Ilo* Renato Mingozzi and *M Ilo* Ferruccio Vignoli, all members of 1s *Sq*. Its task was fly as high cover and patrol the combat areas over which other *ANR* fighters were operating. By providing timely early warning of the presence of Allied fighters, the *Occhio di Lince* sections prevented their comrades from being surprised on a number of occasions and they continued to carry out this important duty until war's end.

Having moved base from Cascina Vaga to Villafranca (Verona) on 2 July, *Io Gruppo Caccia* continued to achieve small successes – and to lose some of its own, albeit in similar small numbers. Nine *Gustavs* were airborne on 12 July, again to attack A-20s and P-47s. *Cap* Carlo Miani (*Nucleo Comando*) and *M Ilo* Giuseppe Desideri of 3a *Sq* each claimed one A-20 while on the debit side *Ten* Alfredo Fissore of 1a *Sq* was lost when his Bf 109 dived straight into the Adriatic, due it was believed, to him blacking out as a result of oxygen failure.

On 13 July the target was again Havoc formations attacking the Po River bridges. Splitting their force, the Italians released their drop tanks, 2a *Sq* climbing to provide top cover. *Ten* Ugo Drago meanwhile led 1a *Sq* in to the attack. Despite the A-20s maintaining formation Drago latched on to the 'tail end Charlie' and peppered it with fire. The A-20 gradually lost height and finally crashed in the Porto Tolle area near the river mouth. *Serg* Leo Talin claimed a second A-20. Tangling with the Spitfire escort, *Serg Magg* Cavagliano claimed one shot down, the British fighters in turn shooting down *Serg Magg* Luigi Santuccio, who was killed.

Targets for the *ANR Gustavs* continued to be light bombers but in mid-July a number of sorties were flown against 15th AAF heavies. A B-24 was claimed on 20 July with another the following day. The almost inevitable price was the loss of a pilot, *Serg Magg* Luigi Feliciani, on that occasion.

Better success against the American heavies was achieved when 18 II₀ *Grupo Gustavs* flew to Tulin airfield on 24 July to operate under the command of *JG* 53 in the defence of Vienna. For the loss of one Italian pilot, this combined force claimed the destruction of eight bombers and a P-38, this last falling to *Serg* Luigi Mario of 1a *Sq*.

After further combat with Thunderbolts and Spitfires, July ended with one more Italian pilot killed and one made prisoner by partisans. He was later released. Early in August II₀ *Gr* welcomed a number of new pilots fresh from the German training school at Leignitz and on the 8th *Serg Magg* Attilio Sanson managed to extricate himself from a dangerous situation. Having scrambled to intercept a B-26 Marauder formation over the Rimini area, Sanson became surrounded by Spitfires. His Bf 109 having taken hits, the Italian was being hotly pursued by a number of British fighters. Thinking quickly, Sanson suddenly closed the throttle and lowered his flaps, causing the Spits to overshoot. Taking his chance, Sanson opened fire and saw one of his pursuers fall away in flames. He managed to nurse the *Gustav* back to Villafranca where the groundcrew amused themselves by counting holes . . .

German coup attempt

Then on 25 August, the *ANR* was almost destroyed without any enemy action. Hitler had apparently decided to substitute the nationalist air force with an Italian legion, under which pilots would be absorbed into the *Luftwaffe*. In what amounted to a coup, the Germans offered the Italians the chance to 'volunteer' for the *Luftwaffe* or be drafted into *Flak* divisions. Not only did the Italians dislike this dubious offer – they flatly refused to comply with it. The whole affair was ill-timed and badly mismanaged, considering the proven loyalty of the *ANR* to the fascist cause – but a beleaguered Hitler saw it as an insurance that enabled him to withdraw the *Luftwaffe* out of Italy for good. Even though there were few other recriminations over the 'revolt', the Germans ordered the withdrawal of all II₀ *Gr*'s Bf 109s to Germany, leaving the unit with nothing to fly.

By the end of September the situation was resolved; *Gen* von Richthofen who had engineered the coup attempt under the direct order of the *Führer*, was recalled to be replaced by *Gen* von Pohl. The latter arrived in Italy with new orders which would not only ensure a future for the *ANR* but increase its potential strength. I₀ and II₀ Gr C would be trained in Germany on the Bf 109 and II₀ *Gr* would collect new Bf 109s from Thiene (Vicenza). This arrangement suited the Italian pilots admirably but some six weeks' operations had been lost during which the Allied air forces had inevitably grown stronger.

It was mid-October before II₀ *Gr* was again fully equipped with a mixture of new machines and those left behind by I./*JG* 4 and III./*JG* 53 when they withdrew from Italy. The hiatus – and the responsibility of the defence of Northern Italy – had bred a desire in the pilots to prove their worth in the continuing struggle. On 18 October Ia *Sq* occupied a new base at Ghedi (Brescia) with the other II₀ *Gr Squadriglia* remaining at Villafranca. On the 19th the *Grupo* was once again declared operational.

Luck attended the *ANR*'s return to combat for an interception call in the afternoon of 19 October materialised as an unescorted formation of 319th BG B-26s. Twenty Messerschmitts from all three *Squadriglia* prepared for action under the command of *Cap* Miani. The recent non appearance of enemy fighters in strength over Northern Italy may have bred a false sense of security in some Allied units and these particular B-26s of the 440th BS apparently carried no nose or waist guns. Three B-26s were definitely shot down by the Bf 109s, the pilots of which overheard radio calls from the Marauder crews conforming that they were being attacked by Italian aircraft. This engagement was not without loss as *Serg* Leo Talin crashed while attempting to land his damaged *Gustav* a few kilometres from Villafranca. On the last day of October 1944 a clash between 25 Bf 109s and P-47s of the 364th FS, 340th Group resulted in the loss of three *Gustavs*, two pilots being killed and one baling out. One P-47 was shot down.

Allied plans now turned on breaking the stalemated ground war in front of the Gothic Line by bombing the rail lines through the Brenner Pass to cut off all German supplies. Attacks were also made on the northern Italian airfields, among them Villafranca, but without damaging many aircraft. II₀ *Gr* nevertheless felt it prudent to change bases and a move to Aviano was made early in November. It was from there on the 4th that six *Gustavs* sortied against a B-17 formation and claimed three downed. The target was again Marauders on the following day, the 320th Group being on the receiving end of the Italian attack. Head-on passes destroyed three B-26s.

Combat with American mediums occupied succeeding days and on 10 November *Ser Magg* Baldi was lucky to escape with his life. Preoccupied with watching a crippled B-26 fall, the Italian pilot failed to notice what was creeping up on him. Suddenly the red spinner of a P-51D appeared right behind and Baldi firewalled his throttle to escape. The Mustang had not previously been encountered by the Italians, but its reputation was familiar enough. Yet Baldi manoeuvred his *Gustav* with *élan*, keeping low over the tree-lined mountain passes until the American abandoned the chase. Baldi, realising his aircraft had a damaged gear leg as he came in to land, held the Bf 109 level until it lost enough speed then gently settled the starboard wing.

Elsewhere on 10 November 3a *Sq* lost *Serg Magg* Pacini who was cornered by P-47s of the

57th Group before the *ANR* Messerschmitts could reach their target, another group of Marauders. Pacini was killed when his *Gustav* succumbed to the combined fire of a number of Thunderbolts.

By early November the influx of new fighters from Germany had included a few examples of the potent G-14 model and with the completion of flight training, II₀ *Gr* had an overall strength of about 40 aircraft. Serviceability was high and a reserve of 20 Bf 109s was available, the G-6 continuing to be the predominant model. There was little doubt that the Italian Bf 109 force represented a danger to Allied bomber crews even though its efforts could only realistically achieve modest results against such large, well protected formations covered by swarms of escort fighters. However small the opposition might have been, Allied planners sought total air superiority and to achieve it, they continued to bomb the main *ANR* airfields. Consequently on three days in November, 15th AAF B-24s plastered Villafranca, Osoppo and Aviano. For all the weight of bombs, the toll of destroyed Bf 109s was a modest seven with eleven damaged.

Deteriorating weather conditions brought a slowdown in both ground and air fighting as the winter of 1944-45 swept in to create records for severity. Initiating a system of patrols over the main roads of Lombardia to spot Allied transport movements, II₀ *Gr*'s Bf 109s generally found little action. On 10 December a B-25 formation was intercepted, the Bf 109s destroying at least one bomber with two more claimed plus one damaged. Intervening P-47s of the 57th FG resulted in one Bf 109 being shot down, the pilot baling out.

After shooting down a Spitfire on 22 December, the Italians' fortunes underwent a reversal, fortunately without pilot casualties. In two separate ground attacks on Thiene and Villafranca, P-47s destroyed 20 Bf 109s. The year rounded out with victories over three P-51s, one of them falling to *Cap* Drago. Thereafter the weather clamped down and II₀ *Gr* was effectively grounded until January 1945.

January also saw the arrival of 1₀ *Gr* from Germany and at that point II₀ *Gr* reorganised by renumbering its *Squadriglia* much on the lines of German *Staffeln*. As II₀ *Gr* had *Squadriglia* numbered 1 to 3 it was logical to introduce the numbers 4-6 for II₀ *Gr*. Bases now included Lonate Pozzolo.

Patrols similar to the German *frei Jagd* sorties were now flown regularly and on 20 January one of these intercepted P-47s. A dogfight claimed the life of *S Ten* Pietro Brini of 5a *Sq* but he was immediately avenged by *Ten* Alberto Volpi flying a Bf 109G-14/U4 who despatched the offending Thunderbolt with a few 30-mm cannon rounds.

New Bf 109s from Germany more than made good any attrition suffered by the *ANR* and late January saw the arrival of examples of the G-10, G-10AS and G-10/U4. II₀ *Gr* C was the initial recipient of these machines and both 4a and 5a *Sq* put them to good use when OKL requested take off and landing cover for the Me 262s of KG(J) 6 using Osoppo airfield on a temporary basis during February and March.

A large scale air battle developed over Vicenza on 4 February when the Italian *Gustavs* again clashed with the 57th FG. Two P-47s were shot down. More successes and losses, mostly in the small numbers that had become normal for both sides in this war theatre, marked the next few weeks. Three pilots were killed with seven Bf 109s lost while claims for a number of Allied bombers and fighters were filed. Interceptions could be frustratingly inconclusive for the Italian fighter pilots; numerous US bombers, attacked repeatedly, could be seen to be badly damaged, yet they continued to fly on.

With the spring of 1945 bringing better flying weather, the Axis position deteriorated further. The *ANR* looked however to be in excellent shape; its Bf 109 *Squadriglia* pilots flew sorties with considerable skill built on experience to show that in capable hands, the late model *Gustav* remained a very effective weapon.

One operational method by the II₀ *Gr* C and which much impressed their German allies was to take fighters off alternately from the four corners of the airfield. This saved considerable time over using only one runway and enabled the unit to get nine Bf 109s airborne within a few minutes.

Despite Germany's predicament, supplies of fighters to Northern Italy did not falter and between 13 and 14 March 5a *Sq* of II₀ *Gr* C received nine G-14/AS and four G-14 models, enough to re-equip the entire unit. In company with 4a *Sq*, 5a *Sq*'s new Messerschmitts were deployed on a new and highly responsible duty during mid-March, namely to fly take off/landing cover patrols for three Ar 234 reconnaissance bombers based at Udine. One such sortie on 23 March saw seven Bf 109Gs of 4a *Sq* protecting an Arado from the unwelcome attention of P-47s of the 79th Group. One Thunderbolt went down and a second was claimed. Later on 23 March 4a *Sq* shot down a B-25 near Aviano.

On 22 March a German report on II$_0$ *Gr* C strength listed 16 Bf 109G-6s plus a total of 30 G-14, G-14/AS, G-10/AS and G-10 variants. Pilot strength was 43 officers and NCOs – but the most significant factor was the low rate of pilot fatalities, 4a *Sq* having lost only two between 30 May 1944 and 23 March 1945, together with an admirably low rate of aircraft attrition.

Then disaster struck. On 2 April 24 *ANR* Bf 109s, eight from all three II$_0$ *Gr Squadriglia* plus three from Nucleo Comando, scrambled to intercept a large formation of medium bombers with heavy fighter escort reported over the Alps en route from Corsica to their Italian targets. Three Messerschmitts were obliged to abort due to

Scenes like this were common on airfields all over Germany during the summer of 1945, and near Hamburg the skeletal remains of a Bf 109G-12 were found among the single-seat fighters. *(via D Howley)*

engine trouble but the rest, led by *Magg* Miani, pressed on towards South Venito, aiming to avoid the American escort. Then Miani himself had to land with his radio inoperable and *Magg* Bellagambi, leading the 5a *Sq* section, had his vision obscured by a windscreen that became covered in oil. Meanwhile, 4a and 6a *Squadriglia* attacked the Mitchells.

Just as 4a *Sq* completed its first pass, the escort fell on the Bf 109s. Pilots of the 350th FG achieved surprise and shot down three *Gustavs*. Then all hell broke loose as the Italians plunged into the mêlée. A fourth Bf 109 went down, *Ten* Giorio being killed. *Ten* Bruno Betti and *Ten* Alessandro Abba were also lost.

Climbing fast to 28,000 ft, *Capt* Drago's aircraft gradually got ahead of the P-47s which were also climbing. Levelling out, Drago had just initiated a dive to cut off the Thunderbolts when Spitfires bounced the Messerschmitts. They caught *Serg Magg* Baldi's aircraft, which fell away, its tail torn off. Baldi got out of his stricken fighter despite a torn parachute canopy.

Further misfortune was to follow even after the battle was supposedly over. Heading back to Villafranca, 6a *Sq* in company with some 5a *Sq* machines, received clearance for a 'safe' landing. A group of P-47s then jumped the Italians at their most vulnerable, lining up to land. *Serg* Mario Archidiacono was killed when the P-47s opened fire, this action bringing a response from the local airfield defences. In the mêlée, friendly *Flak* hit one Bf 109 – although the pilot was able to bale out – and other aircraft belly-landed.

A Thunderbolt shot down *Ten* Piolanti when 4a *Sq* returned to Aviano and another Bf 109 was to fall before nightfall brought a disastrous day to a close. No less than seven II$_0$ *Gr* C pilots had been killed and 14 aircraft destroyed. Undaunted and with replacement aircraft (including two Bf 109K-4s) having been collected from the Maniago *Luftpark* (aircraft park) during the first week of April, the *ANR* fought on. The *Luftwaffe* Order of Battle for 9 April included II$_0$ *Gr* C with 34 Bf 109s on hand, only 16 of which were serviceable.

A final US Army offensive began on 14 April with troops breaking through to the north Italian plain under a huge air umbrella. On 19 April the

Italian Bf 109s put up something of a 'maximum effort' with 21 fighters. The two Bf 109K-4s, part of 6a *Sq*, were among the day's four casualties. The Mustang cover, provided by the 325th Group, had proved overwhelming.

Only a few more flights were to come, mostly of a non-combat nature. German units, notably *Kommando* Sommer and II./*NAG* 11 (and possibly *NAG* 6) made arrangements to fly the Italian Bf 109s to Germany. It was in fact a German pilot who scored the last kill in an *ANR* Messerschmitt when eight *Gustavs* scrambled from Campo-formido on 28 April. In action against P-47s of the 350th Group, one of the latter fell in flames before both formations disengaged and went their separate ways. For the Bf 109 pilots, this was not a return to the Italian airfield but a destination in Germany.

All German forces (and concurrently all remaining Italian units on the Axis side) still occupying northern areas of Italy, surrendered on 29 April. For most II₀ *Gr* C fighter pilots it was with no small relief that they were allowed to return home. But others fell victim to reprisals and were summarily executed for their past allegiances.

The record of the Italian pilots flying with the *ANR* had been exemplary and had in total caused the Allies the loss of fractionally more than one aircraft for every Bf 109 II₀ *Grupo* had lost – a respective ratio of about 99 to 89. The cost to the Italians was 36 pilots killed.

Czechoslovakia became the centre for rebuilt Bf 109s but in mid-1945 airfields still held many intact *Luftwaffe* aircraft, including G-14 Wr Nr 464463 'White 7' in the foreground. Second in line is a G-14 of II./*JG* 4. (Crow)

Chapter 11
Defeat

Under cover of the militarily-stalemated winter of 1944-45 the *Luftwaffe* High Command had managed to preserve a substantial force of fighters for a 'last throw' operation against an Allied target. Actually the remnants of the force that had its origins in Adolf Galland's 'Big Punch', a mass assault on a US bomber formation designed to bring down upwards of 400 bombers at a stroke, it now received new orders. By late December, the picture had changed; with von Runstedt's tanks having taken thinly-held ground positions in the Ardennes by surprise and thrust forward to create a dangerous bulge in the Allied front line, the stage seemed set for an unexpected lengthening of the war. If the Germans could reach the port of Antwerp, the ultimate goal of the *Panzers*, the Allied armies would be cut in two – although there is little indication of how this force proposed to hold off the inevitable, overwhelming Allied counter attack to regain all lost territory.

Nevertheless, the appalling overcast weather enabled the last great German offensive of the war in the West to proceed unmolested by Allied air attack and the *Luftwaffe* fighter force also enjoyed the respite that such conditions allowed. Hitler had for some time doubted the ability of the *Jagdwaffe* to carry out effectively the kind of mass destruction of enemy bombers as outlined by Galland and the early success of the Ardennes offensive brought about an alternative plan to attack Allied airbases in Belgium and Holland – but Operation *Bodenplatte*, like the ground offensive, needed more than luck and bad weather.

A Hungarian *Jagdgruppe* 101 Bf 109G 'Black 12' dumped on an Me 262A. (*Merle Olmsted*)

With the minimum of briefing held at the last possible moment to avoid any security leaks, the weathermen confirmed that on 1 January 1945 conditions would be favourable for a mass strafing attack on Allied tactical fighter and bomber bases in Holland and Belgium. The operational plan listed 18 bases as the main targets, with three as secondaries. Ju 88 pathfinders would lead the force to compensate for the poor navigational skills of some of the fighter pilots.

Ten *Jagdgeschwadern* with a total of 33 *Gruppen* brought some 900 fighters to readiness in the early hours of 1 January. Comprising mostly units flying late model Messerschmitts and Focke-Wulfs, the force had the following 19 Bf 109 units: III./JG 1 (Bf 109G-14); II./JG 2 (G-14/K-4); I and III./JG 3 (G-10, G-14 and K-4); I and IV./JG 4 (G-14 and K-4); III./JG 6 (G-10 and G-14); III./JG 11 (G-14 and K-4); III./JG 26 (G-14 and K-4); all four *Gruppen* of JG 27, only I *Gruppe* of which had examples of the K-4; all three *Gruppen* of JG 53, II *Gruppe* flying some K-4s, and all three *Gruppen* of JG 77 with K-4s and G-14s.

When the anticipated codeword *Hermann* was passed to all airfields, ordering an 0920 attack time, ground mist hampered and delayed the take off, particularly for those units in northwestern Germany. This delay was to prove disastrous, for the *Flak* batteries had been briefed to hold their fire only until the attack force passed over their sector at a set time. Many aircraft flying together at any other time had to be assumed to be Allied – and so it occurred that friendly fire accounted for about 100 German fighters because nobody thought to pass revised timings to the gunners.

Surprise was achieved at most of the bases actually attacked by the fighter force and substantial damage was done – but for the Luftwaffe, *Bodenplatte* was a tactical failure. The loss of about 20 pilots and some 500 aircraft was hardly a disaster to the Allies at that stage of the war.

On the other hand, *Bodenplatte* cost the Germans 237 pilots killed, missing or made prisoner at a time when every pilot even basically trained to fly a fighter was worth his weight in gold. Nevertheless, by attacking airfields *en masse* instead of a bomber formation, the *Jagdwaffe* probably lost fewer pilots.

That said, to put together an attack force approaching 1,000 aircraft was a considerable achievement under the circumstances and one that indeed caught the Allies off guard. The operation might have been better timed but the Germans had little choice but to launch it while a substantial number of fighters remained serviceable and before the weather improved, bringing the enemy fighter bombers back into the fray in force to inevitably whittle away this carefully marshalled strength. Such a situation had already occurred to some extent even before *Bodenplatte* actually went ahead.

Allied reaction to the New Year's Day attack was swift and the poor results achieved at some targets highlighted the level of inexperience that the *Jagdwaffe* was now obliged to rely on for its first-line pilots. There was no opportunity to follow up *Bodenplatte*, for von Runsted's ground offensive made slow progress in the face of the adverse weather and Allied reaction. Pinning down the Allies at Bastogne took precious time and although the situation looked critical, Patton's relief force reached the besieged town on 3 January. By the 9th, the Germans had no choice but to initiate a withdrawal.

Neither was the US strategic bomber offensive overly hampered by the inclement weather. The 8th Air Force had put 850 B-17s and B-24s over oil and communications targets on 1 January, the force being escorted by elements drawn from all its component fighter groups. These latter claimed a total of 17 German aircraft shot down in return for eight bombers and two fighters, loss figures that might have been worse for the Germans if the 15th Air Force had been able to mount a mission. But the 15th's bombers were well and truly grounded by the bad weather which was to plague the force in the closing months of the war.

Bodenplatte had been possible only because German airspace had been free of Allied tactical fighters for weeks, but when the skies over the continent cleared, the stalled Allied air offensive brought a further deluge of fire. Having been caught napping, the USAAF and RAF made every effort to ensure that nothing like it occurred again.

On 3 February the USAAF despatched 1,000

bombers to Berlin while Patton's advance was approaching the west bank of the Rhine. The Russians were on the Oder by that same date and the Germans flung in some of their last reserves to plug a gap in the line around Budapest. It was to no avail.

Even with the German jet fighter programme given top priority, the Messerschmitt plants continued Bf 109 production at an ever-increasing pace. Despite the main and subsidiary assembly plants associated with the Bf 109 and Fw 190 long having been priority targets for the 8th and 15th Air Force bombers, output was not only maintained, but improved upon. Largely due to the efforts of Albert Speer's armaments ministry, many plants had continued to function and timely dispersal of vital component manufacturing into areas almost impossible to detect, let alone bomb, had enabled aircraft output, particularly of the Bf 109, to rise steadily.

Vital as the Me 262 programme was – the only really practical jet along with the Arado 234 – the high command knew that to terminate prematurely conventional fighter production could be courting disaster and on 20 March 1945, *Oberst* Gordon Gollob, *General der Jagdflieger*, issued an ambitious restructuring programme for the fighter arm. This, while estimated to have taken some months to implement fully, was to have had all units then equipped with Bf 109G models convert to the K-4, these complementing the Me 262 and Fw 190, the latter including the Ta 152 derivative, and the He 162. Ultimately, all the Bf 109 units would have converted to turbojet fighters.

Despite most of the Rhine bridges having been demolished, 7 March saw the Allies make an initial crossing at Remargen, where a single track railway bridge had remained intact. In ten days, during which the *Luftwaffe* made repeated attempts to destroy the bridge, the entire west bank of Germany's principal river artery had been lost. Montgomery crossed the Rhine at Wesel on the 23rd while the *Luftwaffe* was also hard put to disrupt the Operation VARSITY airborne assault which opened up the way to the Ruhr. In the south, seemingly little could be done to prevent Patton's Third Army from charging headlong into Mainz, Frankfurt and by month's end, Mannheim.

Above the shrinking territory of the Third Reich, the *Jagdwaffe* fought on, the piston-engined fighters operating alongside the Me 262; in general the results of air combat were disappointing, with aerial victories being increasingly harder to achieve. Extreme frustration and feverish work to overcome the technical difficulties plaguing the Me 262 could do nothing to make what was potentially the world's best interceptor fighter into a more effective weapon.

The *Luftwaffe*'s daily loss records department methodically catalogued the results of prolonging the agony and among the statistics was the fact that between 2 November 1944 and 24 March 1945, eight of the *Jagdgruppen* equipped with the Bf 109 on Defence of the *Reich* duties lost at least 232 Bf 109K-4s, although this figure is probably too low due to the fact that individual aircraft frequently appear as 'Bf 109 G/K' – a comment on the confusion that existed over the various subtle changes between these two late-war models.

Meanwhile the Red Army was similarly unstoppable in its advance on Germany from the East. With Warsaw, Modlin and Cracow taken by 19 January, there were too few German forces to counter successfully the Russian drive into other areas of Poland, including Gydnia and Danzig, both of which were in Russian hands by 29 February. The way to Austria was now open and by the 30th, Russian tanks had crossed the border.

Clearly the *Jagdwaffe* in its present state was powerless to change the military situation and as long as the American bombers remained the primary fighter target, little effort could be spared to support the ground forces' desperate struggle against overwhelming American and Russian forces. Some escort sorties were flown however for the few ground attack aircraft that remained operational, but Allied pilots began to realise just how desperate the Germans were when in those last weeks, Ju 87s were encountered in some numbers over the front lines. If caught by Mustangs, the old *Stukas* suffered terrible losses, even if they had plenty of protection from Bf 109s or Fw 190s.

When the Allied armies achieved a link-up at

A Bf 109G of an unknown II *Gruppe* with the late-war nose band applied to many *Luftwaffe* aircraft, not only fighters. Also of note is a masked-off diamond shape, almost certainly for a *Staffel* badge. Next in line is a Bf 109K-4. *(Olmsted)*

Paderborn on 1 March, the Rhine encirclement was all but complete. The Elbe was in Allied hands by the 12th and enemy guns, tanks and troops were then 60 miles from Berlin. Nuremburg, the cradle of Nazism, was lost on the 20th and although General George Patton was urging his superiors to allow American tanks to capture Berlin, the German capital was left to the Russians. Patton did not pause and had arrived at the Czechoslovak frontier by the 23rd. Two days later, American and Russian troops shook hands at Targau, north-east of Leipzig.

When even the *Jagdwaffe*'s primary target, the USAAF heavy bombers, could no longer be brought down in any numbers, no matter how many conventional interceptors managed to get airborne, it was clear to the *Luftwaffe* high command that more drastic measures were called for. Various ideas were put forward to destroy substantial numbers of bombers at one go, even at this, the 'eleventh hour'. One that appeared feasible was suggested by *Oberst* Hajo Herrmann, father of the *Wilde Sau* night fighter concept.

Herrmann's plan was terrible in its simplicity – volunteer pilots flying Bf 109s would bring down enemy bombers by ramming. It was a grim prospect for the participants and one that reflected the desperation of the Japanese *kamikazes*. No longer would ill-trained pilots have to select a target and make a conventional attack – they would simply hurl themselves into the American formations and smash the bombers out of the sky.

Spurred on by propaganda that in no uncertain terms stressed the vital need for such sacrifice, the response to Herrmann's call for pilots was surprisingly enthusiastic. Some 2,000 men, drawn mainly from *Luftwaffe* training schools, joined the *Rammkommando*. By the first week of April, a force of over 100 Messerschmitts had been assembled for use by the unit, since renamed *Sonderkommando Elbe*. Each Bf 109 was stripped of all unnecessary weight, most fighters having their ammunition load halved.

Although Herrmann wanted three times the number of fighters for the initial operation, this was scheduled to take place on 7 April with what was available, under the code-name *Wehrwulf*. That day the AAF sent 1,314 B-17s and B-24s over Germany along with a fighter escort totalling 898 aircraft. In support of the *Elbe* ramming force were jets and conventional fighters flown by pilots who had every intention of returning home after combat. Jet sorties for the day in fact totalled 59, the highest ever recorded in one 24-hour period.

Operation *Wehrwulf* was a fiasco. Taking off and climbing in ragged groups rather than the mass formation that had been intended, the Bf 109s failed to concentrate. Only eight heavy bombers were believed to have been brought down by physical contact, with a further 15 damaged. American aerial victories reached 64, the victims assumedly being mostly from the ranks of *Sonderkommando Elbe*, as the only other German loss that day was a single Me 262 of *JG* 7 which collided with a B-24. Herrmann's patriotic young pilots, few of whom had any idea of how to combat a Mustang or had the full means to do so, were picked off piecemeal.

These young fanatics had apparently been told on good authority that if they hit a B-17 forward of the vertical fin, the impact of their wing would be enough to break the bomber in two, whereupon they could safely bale out. The reality could be somewhat different. The demise of one of Herrmann's valiant flock was literally felt by the crew of 1st Lt Carrol Cagle's 490th BG B-17, which came under a diving attack from 'eight o'clock high'. The Bf 109 did not avoid the Fortress but impacted in the waist area with its starboard wing, which broke up. The Messerschmitt cartwheeled away, scraping along the bomber's underside before disintegrating. No further ramming sorties were flown.

Operations by the US heavy bomber groups reached a crescendo by mid-April 1945 as with Allied troops having secured virtually all of Germany and her occupied territories, there were few worthwhile targets left. The 15th AAF flew some of its final sorties against tactical targets and on 15 April virtually its entire force of heavies, amounting to 1,235 Fortresses and Liberators, supported by 586 fighters, was despatched to attack strongpoints and troop concentrations in and around Bologna.

If Hajo Herrmann's *Sonderkommando Elbe* demonstrated just how desperate the situation had become for the *Luftwaffe*, it was not unique in attempting to redress the balance by tactics requiring the ultimate sacrifice from pilots. On 16 April Bf 109Gs of II./*JG* 4 escorted *Mistel* combinations from *KG* 200 to attack Red Army units crossing the Oder and Neisse rivers. Tucked into the formation were ten Messerschmitts of *SG* 104, whose pilots had volunteered to carry out suicide attacks on the bridges. Each Bf 109 carried a bomb which was wired to explode when the aircraft impacted the target.

This effort, believed to be the only other occasion when the Germans resorted to deliberate suicide attack, hardly caused the Russians to pause in their crossing of both rivers, an operation completed by the 17th. This and subsequent *Mistel* attacks against the Soviet bridgehead on the Oder achieved very little.

Despite all the carnage and unbelievable confusion, the *Luftwaffe* administrative staff, to its credit, continued to raise the necessary paperwork for awards to be made to outstanding pilots. The final presentation of the Oak Leaves for example, was made to *Fahrenjunker Ofw* Heinz Marquardt of IV./*JG* 51 on 17 April. He was also the last *Jagdflieger* to score 100 victories.

Heavy bombers of both the 15th and 8th Air Forces flew their last missions to Germany on 25 April, the latter's effort amounting to 589

Liberator and Fortress sorties, covered by 584 Mustangs. The bombers struck Pilsen airfield, the Skoda works and marshalling yards in Czechoslovakia as well as German rail centres.

Having genuinely assumed that Hitler himself had lost 'freedom of action' as he put it in a telegram to the *Führerbunker* in Berlin on 23 April, Hermann Göring made himself ready to take over the remains of the Third *Reich*. Typically, Hitler saw this suggestion – based on a June 1941 agreement between the two of them – as treason. Göring was promptly placed under house arrest and the *Luftwaffe* had a new C-in-C, *Generaloberst* Robert Ritter von Greim, effective 24 April 1945.

It will be recalled that it was von Greim who carried out some of the earliest test flights of the Bf 109 prototypes some ten years previously – although at the time of his promotion, that significant event must have seemed like an entire lifetime in the past. Rarely can a general have been given a command at a less opportune time, nor been faced with a force of such potentially enormous power destroyed as much as by the policies of its leaders as its all-powerful enemies. And von Greim's command was brief indeed. On 26 April in response to a *Führer* summons, he was in a Fieseler *Storch* (Stork) en route to a makeshift airstrip near the Berlin bunker. The *Storch* was hit by Soviet AA fire and the new *Luftwaffe* chief was wounded in the foot. He was thus destined to see out the remaining days of his command in some pain but succeeded in leaving the doomed capital and flying to Rechlin on the very day that Hitler took his own life. After 29 April 1945, all need for any further sacrifice on the part of *Luftwaffe* airmen had been removed.

But in those chaotic, closing weeks of the war, a fatalistic feeling prevailed that at least the fight was now for Germany, rather than a regime that few had actively supported. The momentum that the bitter defence of the country had engendered could not simply be switched off and until someone in authority ordered the fighters not to fly, their war went on.

One of Hitler's last orders was that the *Luftwaffe* should mount a strong air attack to help ground forces make one last effort to break the siege of Berlin. The force was to be under von Greim's orders and directed from Rechlin. It was carried out, but by that time the Russian pressure on Berlin was so strong that hundreds more sorties than were then possible by the *Luftwaffe* would have been required to change the inevitable outcome.

By May 1945 the *Jagdwaffe* was still operating from airfields in southern Germany, Austria, Czechoslovakia and Bulgaria. Further north in Courland, the territory that bulges into the Baltic region and which had been bypassed by the main westward thrust of the Red Army, German fighter operations also continued.

Pilots of JG 52, hitherto used to meeting only Russian aircraft in combat, now came up against Mustangs over Romania. To the *Experten*, Erich Hartmann included, the difference was academic. In a series of combats with the fighter escort to 15th AAF bombers, he had shot down seven P-51s to add to his already impressive total of Russian kills – ample proof, if such were needed, that the Bf 109G with an exceptional pilot at the controls, remained a formidable weapon.

JG 51, 52, 53 and 77 rotated between the remaining airfields, component *Gruppen* of all these units continuing to fly a mix of Bf 109G-6, G-10, G-14 and K-4 models. JG 52 was still the leading *Geschwader* in terms of aerial victories, having recorded its 10,000th kill on 2 September 1944 when *Hptm* Adolf Borchers brought his personal score to 118.

Fighter versus fighter combat occurred until the last day of the war, although many Allied sorties were now flown without a sight of *Luftwaffe* aircraft, at least in the air. In some *Jagdgeschwadern*, the fuel shortage had forced a drastic curtailment of operations and the inexperience of the remaining fighter pilots, often as a direct result of them being unable to undertake enough training flights, led to their sudden loss a short time after posting to an operational unit. Other units did not suffer as acutely from a lack of fuel and managed to maintain a regular sortie rate.

Bf 109s continued to fall to roving patrols of RAF fighters, particularly the potent Hawker Tempest. In April 1945 Nos 3 and 486 Squadrons shot down single Bf 109s on three separate days and on 1 May, the latter squadron destroyed what is believed to have been the last Bf 109 claimed by a pilot flying this type. The German machine fell over Bad Segeberg and may have been the aircraft flown by *Ofw* Otto Buss of 14./*JG* 52. (14th *Staffel* of *JG* 52).

Erich Hartmann, now way ahead of his contemporaries, fought on. Promoted to *Major* in February 1945, this outstanding exponent of the Bf 109 shot down one of the last enemy aircraft to fall to the *Jagdwaffe*. Patrolling over Brun, Czechoslovakia on 8 May, Hartmann surprised a Yak-11. Hostilities had but hours to run and when he landed to record the kill, his total had reached 352. In terms of enemy aircraft destroyed, this figure made Erich Hartmann the most successful war pilot of all time. Throughout his combat career, in common with the rest of *JG* 52, he had flown only the Bf 109.

The German surrender on 8 May saw personnel stationed on former *Luftwaffe* airfields, now crowded with American and RAF aircraft, witnessing the arrival of German aircraft with no further hostile intent. Individual *Jagdflieger* accommodated their groundcrews in single-seat fighters for an occasionally hair-raising final flight to the West or to neutral Sweden or Switzerland – anywhere to avoid falling into the hands of the Russians. Some *Luftwaffe* diehards ordered all serviceable aircraft to be destroyed to prevent them falling into enemy hands and airfields reverberated to the explosions of demolition charges. But thousands of aircraft were simply abandoned where they stood.

Aftermath

Post-war analysis showed that production by the Messerschmitt plants had been prodigious. Despite Allied bombing, interdiction of test airfields and dispersal of vital component facilities, the total number of Bf 109s built by all plants was considerably more than 30,000 airframes. Some sources put the actual figure as high as 33,000, which made the Bf 109 almost the most-produced aeroplane in history. It had however to concede first place to the Ilyushin Il-2 (at 34,000) and probably second place as well, to the ubiquitous Po-2, the very widely used two-seat liaison, transport, air ambulance and courier biplane, output of which also reached 33,000.

Being beaten in the production stakes by the Il-2 was in a way fitting, as it had been the *Jagdgeschwadern* flying the Bf 109 which had been partly responsible for a constant flow of replacements being required by the Russian *Schturmovik* units! This was but one aspect of the air war in the East, a voracious war theatre covering enormous distances and seeing some of the heaviest fighting of the war.

It was when the Allies examined German pilot and unit victory lists that the true scope of air combat in this and other theatres became apparent. By flying throughout the conflict, German fighter pilots managed, providing that they beat the odds, to notch up more kills than any other band of pilots on the face of the globe. Their individual scores alone were far in excess of those achieved by entire groups, squadrons or even wings in the Allied air forces, the latter's system of relatively short-duration tours of operational flying for pilots not being followed by the *Luftwaffe*.

Taking the Allied yardstick of five aerial kills (the Germans used a baseline of ten if they followed the practice at all) to make a pilot an ace, no less than 5,000 *Jagdfliegern* had achieved this feat by the war's end. Eric Hartmann's 352 victories was way ahead of any Allied pilot, the Russians coming nearest with Ivan Kohzedub's 65 kills to make him the top *VVS* ace. Gerhard Barkhorn's 301 victories made him the only other member of the '300 club' while Gunther Rall (275), Otto Kittel (267), Walter Nowotny (258), Wilhelm Batz (237) and Erich Rudorffer (222) headed the list of pilots whose score had exceeded 200 kills. Eight others achieved this feat and no less than 82 men scored 100 victories or more.

The statistics of combat can be broken down further to highlight numerous 'firsts' within the *Jagdfliegers'* monumental achievements in combat – but while one might deduce that aerial victories were easier to achieve in Russia than in the West, a few individuals might dispute the fact if it were implied that the risk factor was also better. Georg-Peter Eder, who scored 78, was shot down 17 times and wounded on 14 occasions. Using up fighters in this fashion was far from unusual and most of the *Experten* were obliged to fly multiple examples of the Bf 109 and Fw 190 as a result of combat damage or accidents, quite apart from the delivery of new variants to their unit.

Almost without exception a German fighter pilot who had been in action at the beginning of the war had used the Bf 109 for some of his victories and many men flew no other aircraft type for the duration of hostilities. The greater percentage of aerial kills on all fronts was, understandably enough, scored by the Bf 109 and Fw 190. The Russians admitted the loss of about 45,000 aircraft in aerial combat with German fighters on the Eastern front and the Western Allies put their combined loss figure to German fighters at some 25,000 between 1941 and 1945.

The cost to the Germans was unsurprisingly, enormous: the *Luftwaffe* lost 150,000 men killed, wounded or missing during the war, 70,000 of whom were pilots. But the *Jagdwaffe*'s proportion of this enormous figure was approximately 8,500 killed, 2,700 missing or PoWs and 9,100 wounded.

Chapter 12
Fading away

Along with huge numbers of Bf 109s standing intact on airfields all over Europe, many partially-assembled airframes were discovered, on interrupted production lines and at repair facilities, complete with jigs and tools. Fortunately, not all of these were destroyed in the wholesale scrapping that took place in the chaos immediately following the cessation of hostilities. American and British technical teams earmarked examples of the leading German combat aircraft for shipment and these were duly despatched to the RAE at Farnborough and Wright Field, Ohio for comprehensive flight testing.

Other nations had a more practical use for the Bf 109. In Czechoslovakia, final assembly of the Bf 109G-14 by Avia at Prague-Cakovice was halted when the Germans vacated the country in May 1945. By then, a network of Messerschmitt sub-contractors had been established throughout the country and the new plant owners made an

The infamous Avia S 199 'Mule' of the Czech Air Force failed to live up to its pedigree but it served the Czechs and Israelis when nothing else was available. No 185 is seen in this view. (*via Robertson*)

inventory of existing components and engines prior to restarting production of the fighter for the Czech Air Force.

A shortage of DB 605A-1 engines due to sabotage meant that only 20 G-14s of Czech origin could be completed under the designation S 99 along with two examples of a tandem two-seat trainer based on the Bf 109G-12 and designated the C-10. A considerable re-design was then undertaken to modify the G-14 airframe to take the 1,350 hp Junkers Jumo 211F, ample stocks of which were available.

At maximum output, the DB 605A was rated at 1,474 hp, which meant that the Jumo engine was marginally less powerful; intended to power the Heinkel He 111 bomber, the Jumo required, among other items, a VS 11 airscrew with paddle blades which were considerably broader than any other previously fitted to a Bf 109. The substitute engine had however led to quite substantial changes in the Bf 109's handling characteristics, not the least of which was the fact that torque from the massive airscrew induced a very strong swing during the take-off run. But if the Jumo was far from the ideal fighter engine, there was little alternative but to use it if the Czech Air Force was to acquire a first line aircraft at modest cost. A sizeable number of fighters were required for training pilots to form an embryonic peacetime air defence force and Avia undertook the necessary work to produce what turned out to be the penultimate version of the Messerschmitt Bf 109 line. By so doing, the Czechs extended the original design's lifetime considerably beyond its first 1935-45 decade.

The prototype conversion flew for the first time on 25 March 1947, Avia at Prague-Cakovice starting production of the reworked fighter (as the S 199), later that year and delivering the first aircraft in February 1948. Avia concurrently completed 58 two-seat trainers as the CS 199. By 1949 a total of 422 machines had been completed by the parent firm with the Letov concern at Letnany having established a second production line and built 129, a grand total of 609 aircraft. Most examples of the fighter version carried the standard armament of two fuselage-mounted MG 131 machine guns although an unknown number had a pair of MG 151 cannon installed in the wings outboard of the undercarriage wells. Due perhaps to wing loading limitations, this latter combination generally gave way to externally-mounted cannon.

Avia S 199s entered service alongside the remaining Bf 109G-14s, these latter subsequently being passed to air guard units. The first line Czech Air Force Avia S 199s served until the 1950s and represented the country's first line of defence for at least a half-decade. In Czech service, the *Mezec* (Mule) was found to have some very nasty habits, including a powerful swing on take-off, sluggish acceleration, over-sensitive controls and probably the most demanding landing characteristics of any Bf 109. As soon as possible, the Avias were put up for disposal.

If Czech-assembled Messerschmitt 109s were hardly able to claim the crown of aeronautical achievement in the early 1950s, one nation believed they were almost beyond price – as much as $51,490 each. Put in the context of a new country's total destruction if it could not obtain enough aircraft to defend itself, that was cheap. Israel paid willingly for 25 S 199s and they were duly transferred to serve with the *Chel H'avir* under the code name *Sakin* (Knife).

The Czechs, through no fault of their own, had unfortunately created a mongrel from a thoroughbred with the C 199, a fact that the *Chel H'avir* soon realised. The Mule had nevertheless to be made to perform for the infant state, born on 14 May 1948, was in mortal danger of being smothered by its Arab neighbours long before reaching maturity. Airpower was the key to Israel's survival and pilots and ground crews of No 101 Squadron slaved to get the beast operational. This they achieved, at least in part, the Avia redeeming its investment by a number of morale-boosting aerial victories, enabling the Israelis to 'hold the line' until more capable combat aircraft could be acquired.

Gradually whittled down by attrition, mainly through accidents caused by the vicious take-off swing, the Israeli Avias barely survived until the end of the War of Independence in July 1949. By then they were also operating in the ground attack role alongside Spitfires, in the hands of No 101 Sqn. The Czechs had sold the Israelis ample

Acquiring Bf 109G-6s via a clandestine deal with Hermann Göring, the Swiss air force used the type into the postwar years. Heading this line is 'J-701' with '702' behind it. *(Holmes)*

Uncowled, the Rolls-Royce Merlin of an HA-1112 shows exactly why it can't look very much like a Daimler Benz engine in a genuine Bf 109. *(Lauer)*

stocks of gondola-mounted 20-mm MG 151 cannon and the majority of Avias used in combat were fitted with these guns to supplement the two fuselage mounted MG 131 machine guns. Despite its forgettable flying qualities, the Avia S 199 became something of a symbol of defiance as the first true combat aircraft to bear the Star of David and at least one example was preserved.

Spain's 'fat pigeons'

As one of the first foreign countries in the world to operate the Bf 109, Spain claimed the distinction of building the very last of the line and was responsible for most of the airframes seen currently in various *Luftwaffe* guises, at air shows, in museums and on the screen. Having secured a licence from Messerschmitt in 1943 to manufacture the Bf 109G-2, the contract stipulated that the first 25 airframes (temporarily designated Bf 109J by the *RLM*) would be built in Germany and delivered to Hispano Aviacion.

On arrival these airframes were found to lack propellers, tailplanes and engines and were placed in storage pending delivery of the missing parts. Owing to Germany's worsening military predicament, these did not materialise and the Spaniards sought an alternative powerplant. A 1,300 hp Hispano-Suiza 12Z 89, of conventional V-12 cylinder configuration, was tested in a Bf 109E-1 prior to installation in the G-2s under the designation HA-1109-J1L. Featuring an ugly, drag-inducing radiator, the first Spanish aircraft suffered from a disappointing performance but following the first flight of such a conversion on 10 July 1947, the programme was completed by Hispano.

In the event no HA-1109-J1L entered service with the *Ejercito del Aire* and in order to supply the air arm with a reliable fighter in substantial numbers, it decided to switch the fighter's powerplant to a Hispano-Suiza 12Z 17. This proved to be far superior and as the HA-1109-K1L, the type entered service in 1952. Considered an interim type, the K1L fighter served primarily as a trainer until 1953, when it was decided to install the 1,602 hp Rolls-Royce Merlin 500/45.

The Merlin also introduced a four-blade airscrew on the basic Bf 109 airframe for the first time, enabling the conversion to attain a respectable performance, including a maximum speed of 419 mph and a service ceiling of 33,450 ft. As the HA-1112-M1L alias *'Buchon'* (a kind of deep-breasted pigeon common to the

Passing to Hendon when the RAF Museum was established, Bf 109E-3 Wr Nr 4101, now looks fine in the markings of 2./JG 51, but the years have apparently taken their toll on the interior, making full restoration a daunting task. (*MAP*)

Captured Bf 109s were shipped around the world for evaluation by Allied nations likely to meet it in combat. This G-6 Wr Nr 163824 was at Bankstown, NSW owned by Syd Marshall when this photo was taken in the 1960s.

Seville region), the Merlin installation required a sizeable under-nose radiator and fairing which gave it its nickname. It became the first postwar derivative of the Bf 109 to serve the Spanish air force in substantial numbers, 71 and 72 *Escadrones* (Squadrons) being formed on the type in 1956.

When Seville production was phased out in 1958, 170 Merlin-engined examples had been completed out of a grand total of 239 Bf 109 derivatives that were assembled in Spain. Other units were equipped with both single-seaters and two-seat trainers, for which role the HA-1112-M4L was created by retro-fitting the Merlin to two-seaters which had previously been powered by Hispano engines, a substitution which was also made on single-seaters. Together these Merlin-engined *Buchons* served the Spaniards well until 1967 when they were put up for disposal, 27 M1Ls and one M4L then being eagerly snapped up by the makers of the feature film *Battle of Britain*.

In view of the vast number of Bf 109s built during the war years, very few examples remain today, at least in original form. The majority of airframes masquerading as 109s in the 1990s are survivors of the 40 or so that passed into private or corporate hands after disposal by the Spanish authorities. Enough of these survive to give air show visitors and filmgoers a tangible reminder of the *Luftwaffe*'s most famous fighter and among the type's other film credits have been *Memphis Belle* and the TV series *Piece of Cake*.

Among the notable exceptions to these Rolls-Royce-engined hybrids, most of which are (inevitably) decorated with markings of the

More commonly seen than real Bf 109s these days are Casa HA-1112 *Buchons*. This one was part of the Confederate Air Force when it was based at Harlingen, Texas. (*A F Lauer*)

More German aircraft were earmarked for museum display in the US than anywhere else, but not all made it. One that did was this G-6, seen postwar in spurious markings.

wartime 109 *Experten*, are a handful of survivors of the genuine article. In the UK the ex-8./*JG 77* Bf 109G-2 'Black 6' was unveiled in 1992 after a full 20-year restoration by an enthusiastic team led by Russ Snadden. This aircraft represents not only a living tribute to the thousands of German airmen who gave their lives flying fighters during the war, but an awesome example of the dedication of a small group of people to complete what was, at times, a seemingly insurmountable task.

Parts of other Bf 109s, mainly E models, are on show in major and minor British aeronautical museums. A great many *Emils* fell on English soil or in the surrounding waters during 1940 and although the RAF took pains to remove or destroy the wreckage at the time, enough items have turned up in the last 50 years to provide interested parties in the UK with numerous museum exhibits. Other Bf 109 wrecks have been found almost intact, including the E-4B WrNr 4853, the former property of 2./*JG* 51. After crashing in the Channel off Hythe, Kent on 7 October 1940, most of the airframe was recovered in 1977 by divers working with the Brenzett Aeronautical Museum.

Elsewhere, the *Deutches Museum* in Munich has long held a Bf 109E-3 and the Swiss Transport Museum collection at Dubendorf also includes an example of this sub-type. The RAF Museum at Hendon, London has yet another externally restored Bf 109E-3 WrNr 4101 that force landed at Manston on 14 November 1940 and was subsequently test flown as DG200. After having been through the process of typically unimaginative repainting and the fitting of spurious equipment (including an *Erla Haube*) that is the lot of WWII survivors, 4101 is now a part of Hendon's 'Battle of Britain' collection.

Later 109 versions are currently stored or are on display around the world; Finland's lengthy association with the *Gustav* is reflected by two G-6s at Utti and Jyvaskyla Swedish Air Force bases, there is an F-2/Trop in South Africa and the NASM in Washington D.C. holds a further G-6. There are others, but the world inventory of more or less original Bf 109s from wartime production stood at 16 examples for many years.

The most obvious difference between the postwar Bf 109s and their wartime counterparts is the type of engine fitted. Jumo or Rolls-Royce powerplants completely alter – if not spoil – the lines of the original and it has occurred to some to substitute a Daimler Benz when an engine in serviceable condition can be located. Consequently, Messerschmitt-Bolkow-Blöhm partially restored an HA-1112/ G-6 to flying condition in 1983 by substituting its Merlin engine for a DB 601D. This task is not, apparently, as straightforward as it might at first appear and this particular aircraft was subsequently written off in a crash-landing. The rarity of original German engines is only one of the many challenges faced by anyone in a position to refurbish a surviving Bf 109 airframe but that the effort to do so is worth it is surely beyond doubt.

Increasing access to remote areas of the former Soviet Union and Eastern Europe has yielded priceless relics of the vast air activity over the area during the war years and if more Bf 109s do reach the West, they will almost certainly be from the former Eastern Front. It is fervently hoped that this will happen in due course so that more examples may be placed on display in museums, if not made to fly again. In the early part of the 1990s, this is already happening and at the time of writing (February 1994), a second 'genuine' Bf 109 has flown.

This is Hannes Dittes' restoration based on his Spanish *Buchon* registered D-FEHD and a Bf 109G-10 WrNr 151591, which was discovered in Czechoslovakia in 1992. A machine originally fitted with the taller wooden tail, the restoration features a number of new parts – including the tail – and other items acquired from various sources. The DB 605D-1 engine was a real find in

(Opposite page)
A survivor of a sizeable wartime fleet of Bf 109s captured by the RAF, Wr Nr 4101, (DG200) a Bf 109E-3 was trotted out for numerous displays and at one point acquired an *Erla Haube* in place of its original canopy. It was kept at Biggin Hill in Kent for many years, where this picture was taken.

'Black 6' took to the air again in 1993 for the first time since 1941 after Russ Snadden's team achieved a magnificent restoration to flying condition. (*MoD via R L Ward*)

Turin, Italy, the powerplant, still resting on its transport cradle, having doubled since the war as a wall reinforcement!

Painted to represent an aircraft of *NJG* 11, the aircraft's rudder bears the 30 kill scoreboard of *Maj* Friedrich-Karl Müller, a close friend of Hannes Dittes before his untimely death in 1991. Müller's last wartime aircraft was a Bf 109K-4 'Black 2', the rudder of which he managed to retain. A promise that the restoration would bear Müller's markings was honoured by Dittes and the aircraft was expected to start test flying from Duxford in the hands of Ray Hanna of the Old Flying Machine Company during 1994.

This gradual increase in the number of authentic Bf 109s is gratifying and a tangible reminder to the many surviving German fighter pilots who thus have the chance at reunion gatherings and air shows, to re-familiarise themselves with the machine that carried them through combat. They are paramount among those who would undoubtedly place the Messerschmitt Bf 109 at or near the top of any list in aviation's hypothetical hall of fame as one of the most successful fighter aircraft of all time.

Index

A
Adler Tag ('Eagle Day') 47-8
Aeronautica Militar Espanola 19
Aeronautica Nazionale Repubblicana (ANR) 120-5
Aéronavale 38
Afrika Korps 56, 57, 62, 64, 65, 67, 68
Amiot 143 35
Arado 4, 5, 8
  Ar 80 4, 6
  Ar 234 124
Armée de l'Air 33, 35, 38
Augsburg 1, 7
Avia S 199 *134*, 135-6
Avila 22

B
Balthasar, *Maj* Wilhelm 23, 72
Bar, Heinz 75
Barkhorn, Gerhard 132
Batz, Wilhelm 132
Bauer, Hubert 4
Bayerische Flugzeug Werke (BFW) 1, 2, 3, 4, 5, 6, 9
Beckh, *Maj* Friedrich 86
Berlin 8, 98
Bertram, Otto *Lt* 25, 48
Bf designation *see under* Messerschmitt A.G.
Bloch 155 35
Boeing B-17 Flying Fortress 68, 72, 74, 75, 108
Bolzof, *Lt* Helmut-Felix 26
Bonin, *Oblt* Hubertus von 26
Borchers, *Hptm* Adolf 131
Boulton Paul Defiant 37
Breguet 693 35
Bretnutz, *Lt* Heinz 26, 82
Bristol
  Beaufighter 73
  Blenheim 35
Brucks, *Uffz* Helmut 25
Bulgaria 110
Buss, *Ofw* Otto 132

C
Cavagliano, *Serg Magg* Carlo 118, 120, 121
'Chain Home' 45
Channel Islands 42
Chel H'avir 135-6
Claus, *Oblt* George 70
Co-Belligerent Air Force 120
Condor Legion 9, 19-27
Consolidated B-24 Liberator 74, 75, 108
Crete 63
Croneiss, Theo 1
Curtiss
  Hawk 75 33, 35, 36-7
  P-40B/C Warhawk 63
  P-40D/E Kittyhawk 63, 64
Czech Air Force 134-5

D
Dahl, *Maj* Walter 101, 102
de Havilland Mosquito 77, 103
Desideri, *M Ilo* Giuseppe 121
Dewoitine D 520 35
Dickfeld, *Oberstlt* Adolf 87
Dittes, Hannes 141, 142
Dornier 8
  Do 17 36, 46
Douglas A-20 Boston/Havoc 65, 68, 74, 103
Drago, *Cap* Ugo 118, 120, 121, 123, 124
Duhn, *Fw* Ede 85
Dundas, *P/O* John 70
Dunkirk 37-8

E
Ebener, Kurt 88
Eder, Georg-Peter 75, 132
El Alamein 67
Enigma 45
Ensslen, *Lt* Wilhelm 26
Erla 7-8, 73, 92, 114
Erla Haube 92, 96, 114
Espenlaub, *Obfw* Albert 67

F
Fairey Battle 33, 35
Fiat
  CR 32 23, 24, 26
  G 55 118
Fiby, *Lt* 70
Fieseler 8

Finland 80, 110
Flugzeugbau Messerschmitt Bamberg 1
Focke-Wulf 4, 8
   Fw 159 4, 6
   Fw 190 74, 75-6, 77, 89, 106, 111, *112*, 117, 190-1
   Ta 152 116, 128
Fokker D XXI 35, 38
France, Battle of 34-9
Francke, *Dipl-Ing* Carl 6, 8
Franco, General 19, 20, 22, 26
Franzisket, *Oblt* Ludwig 61

G
Galland, *General* Adolf 24, 36, 49, 70, 72, 81, 101-2, 115, 120
Gentzen, *Hptm* Hannes 30
Gloster Gladiator 35, 38
Gollob, *Oberst* Gordon 82, 86, 128
Göring, Hermann 18, 39, 46, 47, 48, 69, 77, 78, 105, 131
Grabmann, *Hptm* Walter 25
Graf, *Lt* Hermann 54, 63, 85
Graf Zeppelin 53, 54
Greece 62
Greim, *Generaloberst* Robert Ritter von 6, 131
Grommes, *Hptm* Walter 54
Gutbrod, *Oblt* Paul 33, 37

H
Hamburg 77
Handley Page Halifax 72
Handrick, *Hptm* Gothard 24, 25
Harder, *Oblt* Harro 22, 23
Harth, Friedrich 1
Hartmann, *Major* Erich 88, 131, 132
Hawker
   Hurricane 35, 36, 38, 44, 51
   Tempest 131
   Typhoon 73, 103
Heinkel, Ernst 3
Heinkel 4, 5
   Hae 51 19, 20, 21
   He 111 22, 36, 46, 135
   He 112 4, 6, 7, 21
   He 162 128
Held, *Fw* Alfred 32
Herrmann, *Oberst* Haro 77, 129, 130
Herzog, *Fw* Gerhard 37
Hess, Rudolf 2
Heyer, Hans-Joachim 87
Hille, Fritz 2
Hispano Aviacion
   HA-1109-J1L 137
   HA-1109-K1L 137

HA-1112-M1L 137-8
Hitler, Adolf 2, 6, 19, 34, 39, 55, 56, 63, 88, 90, 105, 115, 122, 126
Homuth, *Oblt* Gerhard 61, 67
Honess, *Uffz* Guido 21
Hrabach, *Uffz* Dieter 21
Hungary 111

I
Ihlefeld, Herbert 63
Ilyushin Il-2 85, 132
Israel 135-6
Italy 118-25

J
Joppien, *Hptm* Hermann-Freidrich 85
Junkers
   Ju 52 35, 36, 84
   Ju 88 36, 46, 75, 111
   Ju 87 36, 38, 46, 60-1, 65, *112*, 128

K
Kageneck, *Oblt* Erbo Graf von 65
Kassel 8
Keller, *Hptm* Lother 82
Kiminsky, *Fw* 70
Kittel, Otto 132
Knotzsch, *Flugkapitan* Hans-Dietrich 6
Kohzedub, Ivan 132
Kothmann, *Lt* Willi 61

L
Lavochkin
   La-5 87
   LaGG-3 83, 87, 90
Leie, *Oblt* Erich 70
Liesendahl, *Oblt* Frank 72
Lippert, Wolfgang 24, 67
Lockheed P-38 Lightning 68, 76, 103
London 50, 51
Luftflotte 1 79, 111
Luftflotte 2 43, 47, 49
Luftflotte 3 43, 47
Luftflotte 4 79, 86, 111
Luftflotte 5 111
Luftflotte 6 111
Lufthansa 2
Luftwaffe
   Jagdgeschwader z.b.V. 102
   Jagdgruppe 88 14, 19, 20, 22
   JG 2 34, 37, 50-1, 70, 71, 72, 73, 74, 75, 103, 127
   JG 3 34, 35, 37, 64, 65, 69, 79, 85, 86, 88, 103, 115, 127
   JG 4 88, 108, 127

INDEX

JG 5 87, 103, 108
JG 6 108, 127
JG 11 77, 103, 127
JG 26 34, 35, 38, 51, 56, 63, 67, 70, 71, 72, 74, 115, 127
JG 27 35, 37, 51, 57, 59, 61, 62, 63, 64, 67, 69, 79, 86, 103, 127
JG 51 34, 35, 51, 52, 69, 70, 73, 79, 85, 86, 87, 131
JG 52 33, 36, 37, 51, 63, 64, 79, 85, 86, 87, 88, 131
JG 53 33, 34, 51, 69, 73, 79, 88, 103, 108, 115, 127, 131
JG 54 34, 61, 62, 79, 85
JG 77 32, 33, 37, 53, 54, 62, 63, 64, 65, 69, 72, 79, 81, 86, 87, 88, 115, 127, 131
JG 234 28
JG 300 77, 102
JG 301 77, 103
JG 302 77
LG 2 53, 62, 63, 79
NJGr 10 77
Tragergruppe 186 53
ZG 1(JGr 101) 29, 108
ZG 2 (Jgr 102) 29, 30
ZG 26 29, 51, 79
ZG 26(Jgr 126) 29
ZG 52(JGr 152) 29
ZG 76(JGr 176) 29
Lusser, Robert 3, 4
Lutzow, *Oberst* Gunther 20, 21, 22, 48, 86

M
Maachi
 C 202/205 118
 MC 72 9
Malta 55, 57, 65
Marquardt, Heinz 130
Marseille, Hans-Joachim 59, 65, 66-7
Martin B-26 Marauder 68, 103, 121
Messerschmitat, Wilhelm Emil 1-2, 3, 4, 7, 8, 117
Messerschmitt A.G.
 Bf 109B 7, 8, *8*, 9, *9*, 20, 22, 24, 27, 53
 Bf 109C 10, *11*, 12, 24, 33
 Bf 109D *10*, *11*, 12, *12*, 18, 25, 28, 30
 Bf 109E 12, 13-14, *13*, *16*, *17*, 18, *29*, *30*, 32, 33-4, 38-9, 40-2, *41*, 42, *42*, 44, 45, 51, 52, 53, 54, 56, 57, *57*, 59, 61, 62, 69, 71, 79, 80, 81, *81*, 82, 86
 Bf 109E-1 *15*, 16, *16*, *17*, 18, *28*, *28*, 66, 79, 140
 Bf 109F 43, 52-3, *52*, 59, 60, 64, 65, 70-1, 71, 72, 79, 80, 81, 83, 85, 91, 95, 97, 99, 104, 116
 Bf 109G 67, 69, 73, 74, 75, 77, 78, 89, *89*, 92-3, *93*, *94*, 96-7, 97, 101, *101*, 102, 103, *103*, 104, *105*, *106*, *107*, *109*, *110*, 112-13, *112*, 118-19, 123, 124, *124*, *125*, *126*, *129*, 131, 134-5
 Bf 109H 116
 Bf 109K 113-115, *113*, *114*, 128, *129*

Bf 109T 7, 53-4
Bf 109 V prototypes *3*, *4*, *5*, 6, 7, 8, 9, 20, 53
Bf 109Z *Zwilling* 117
Bf 110 6, 28
Me 209 V5 117
Me 262 116, 128
Messerschmitt-Bolkow-Blöhm 141
Meyer, *Oberstl* Egon 98
Meyer, *Lt* Hans-Karl 24
Miani, *Cap* Carlo 121, 122, 124
Michel-Raulino, Baroness Lilly von 1-2
Milch, Erhard 2-3, 4, 5, 7
Mistel operations 111-12, 130
Mix, Dr Erich 37
Mölders, Werner 23, 24, 25, 26, 33, 37, 49, 70, 81, 84, 85
Montgomery, Field Marshal Bernard (*later* Lord) 65-6, 128
Morane Saulnier MS 406 35
Müller, *Maj* Friedrich-Karl 142
Muncheberg, *Oblt* Joachim 56
museum displays 141

N
Neumann, *Hptm* Eduard 48, 51, 57
nicknames and service slang vii
Nitschke, Gerhard 6-7
Normandy invasion 103-4
North American
 B-25 Mitchell 68, 103
 P-51 Mustang 96, 98, 100-1, 103, 122
Nowotny, *Lt* Walter 85, 132

O
Oesau, Walter 24, 25, 48, 75, 84
Operation *Barbarossa* 63, 79
Operation *Bodenplatte* 126-7
Operation *Cerberus* 72-3
Operation CRUSADER 64-5
Operation DYNAMO 37
Operation HUSKY 119
Operation JUBILEE 73
Operation MALLORY 121
Operation MALLORY MAJOR 121
Operation *Seelöwe* 46
Operation *Taifun* 86
Operation TORCH 67-9
Operation VARSITY 128
Operation *Wehrwulf* 130
Operation *Weserubung* 34
Oschersleben 8

P
Patton, General George 128, 129

Pflanz, *Oblt* Rudi 70
Philipp, Hans 30, 85
Po-2 132
Poland 29-31
Polikarpov
 I-15 9, 20, 21, 22, 24
 I-16 1, 9, 20, 21, 22, 24, 83-4
Pringel, *Maj* Rolf 21, 22, 72
PZL 11 30

R
Rall, Gunther 36, 63, 85, 132
Rammkommando 130
Rechlin 6, 7, 101, 131
Redlich, *Lt* Karl-Wolfgang 26
Redlich, *Oblt* Karl-Heinz 61
Regensburg 7, 73
Reggiane 2002 118
Regia Aeronautica 55, 59, 61, 64, 118-20
Republic P-47 Thunderbolt 76, 98, 103
Restemeier, *Hptm* 9
Richthofen, *General* Wolfram von 20, 26, 122
Ritterkreuz 75, 84
Rommel, Erwin 56, 62, 65, 68
Royal Air Force (RAF)
 Bomber Command 77
 Fighter Command 44-5, 49, 51
Rudorffer, Erich 132
Rumania 69, 110

S
Schellmann, *Hptm* Wolfgang 22, 25, 48, 82
Schiess, *Oblt* Franz 69
Schleif, *Lt* 9
Schlichting, *Oblt* Joachim 22
Schmidt, *Oblt* Johannes 73
Schob, *Uffz* Herbert 26
Schopfel, Gerd 48
'Schrage Musik' 78
Schulz, Otto 65
Seidemann, *Major* Hans 8
Short Stirling 72
Sicily 56, 69, 119
'Sitzkreig' ('Phoney War') 33
Snadden, Russ 141
Sochatzky, *Oblt* Kurt 82, 85
Sonderkommando Elbe 130
South African Air Force (SAAF) 64, 119

Soviet Union 18, 80-90, 109-11, 141
Spain 19-27, 136-8
Sperrle, Hugo 22
Sportflug GmBH 1
Stahlschmidt, Hans-Arnold 67
Stalingrad 88
Steinhoff, Johannes 69, 87
Stor, Willi 18
Supermarine Spitfire 37, 38, 44, 49, 52, 65, 73, 103, 118
Switzerland 18, 78

T
Tidal Wave 69
Trautloft, *Lt* Hannes 9, 20, 48, 49
Troitzsch, *Fw* Hans 32
Tupolev SB-2 21, 23, 24, 25, 26

U
Udet, Ernst 6, 7, 8
Ultra 45
Ursinus, Oskar 1
USAAF
 15th AAF 130
 8th AAF 73-4, 75, 76, 96, 97, 98-9, 127-8, 130-1

V
Vickers Wellington 32
Voennof Vosdushyne Sily (VVS) 80-1, 82-4, 86, 87, 88, 90

W
Warnemunde 8
Warspite 63
Werra, *Hptm* Franz von 40, 82
Western Desert 57-61, 62-3, 64-8
Wick, *Major* Helmut 49, 70
Wiener Neustädter Flugzeugwerke (WNF) 7
Wilde Sau 77
Windermuth, *Uffz* Heinrich 14, 26
Wolf, *Uffz* 70
World Speed Record (1937) 9
Wurster, Hermann 6, 9

Y
Yakovlev
 Yak-1 83, 90
 Yak-7/9 87
 Yak-11 132
Yugoslavia 18, 61-2, 111